Bioinformatics

Bioinformatics

Pamela Scheffler
Maria Guadalupe Vizoso Pinto
Renis Maci

Kruger Brentt
Publishers
2025

Kruger Brentt Publishers UK. LTD.
Company Number 9728962

Regd. Office: 68 St Margarets Road, Edgware, Middlesex HA8 9UU

ISBN: 978-1-78715-443-8

For information on all our publications visit our website at http://krugerbrentt.com/

PREFACE

Bioinformatics plays a crucial role in modern biological research by combining the power of computer science, statistics, and biology. Its significance lies in its ability to analyze vast amounts of biological data, such as genomic sequences, protein structures, and metabolic pathways, to uncover patterns that would otherwise be impossible to detect. With the rapid advancements in sequencing technologies, such as next-generation sequencing (NGS), the field generates massive datasets that need sophisticated computational tools for storage, analysis, and interpretation. Bioinformatics helps make sense of this data, leading to more precise insights into gene function, disease mechanisms, and evolutionary processes. This approach holds the potential to revolutionize healthcare by shifting from a one-size-fits-all model to more targeted and effective treatments. Bioinformatics also aids in the discovery of new drug targets, biomarkers for early disease detection, and the design of novel therapeutic strategies, all of which can have profound impacts on public health. It enables the integration of large-scale data from various omics fields, such as genomics, transcriptomics, proteomics, and metabolomics, to provide a holistic view of biological processes. This integrative approach is crucial for studying complex diseases, understanding microbial communities, and advancing evolutionary biology. The ability to model and simulate biological systems using bioinformatics tools also accelerates the development of synthetic biology and biotechnology applications, fostering innovation in agriculture, environmental science, and bioengineering.

Bioinformatics is a rapidly evolving field that sits at the intersection of biology, computer science, and statistics. It plays an essential role in modern biological research, facilitating the analysis and interpretation of vast amounts of biological data. This book is designed to provide a comprehensive introduction to bioinformatics, covering its foundational concepts, essential tools, and key applications across various areas of biological research. The first section of this book provides readers with a foundational understanding of bioinformatics, beginning with an introduction to key concepts in molecular biology, biological databases, and sequence analysis techniques. In a field where technology and methodologies are advancing at a rapid pace, having a strong grasp of these core areas is essential for both newcomers and experienced researchers alike. By laying a solid groundwork, this section ensures that readers are well-prepared to navigate more advanced topics in the subsequent chapters. The basic principles explored here will be critical for understanding the complexities of biological data, how it is structured, and the various techniques used to analyze it efficiently.

Following this foundational introduction, the book delves into more specialized and advanced topics in bioinformatics. Chapters cover a wide array of subjects, such as genome assembly and annotation, RNA-seq, and transcriptomics—each of which plays a pivotal role in interpreting genomic and transcriptomic data at a deeper level. In addition, the book addresses protein structure analysis, providing tools to examine the relationships between protein sequence, structure, and function. A dedicated section on bioinformatics algorithms explores the computational methods used to extract meaningful insights from biological data. These topics are not only central to bioinformatics research but are also critical in applications ranging from drug discovery to understanding disease mechanisms at a molecular level.

Beyond these technical aspects, the book also explores how bioinformatics is driving innovation in key areas such as personalized medicine, cancer genomics, and systems biology. Readers will gain a deeper understanding of how bioinformatics tools are revolutionizing healthcare by enabling the tailoring of medical treatments based on individual genetic profiles. The role of bioinformatics in cancer genomics—particularly in the identification of genetic mutations and biomarkers—is highlighted as one of the most promising areas in modern medicine. The final chapters provide a forward-looking perspective on how bioinformatics will continue to shape the future of biological research and healthcare, offering insights into its potential to address complex challenges in genetics, disease treatment, and therapeutic development.

We are deeply grateful to all the individuals, as well as the various books, manuals, periodicals, newsletters, magazines, journals, and other resources, that contributed to the preparation of this book. Despite our best efforts, some errors may have inadvertently occurred during the compilation and editing process. We welcome any further queries, constructive suggestions, and criticisms that could help improve the book. Such feedback will be gratefully acknowledged and incorporated in future editions.

Pamela Scheffler

Maria Guadalupe Vizoso Pinto

Renis Maci

CONTENTS

1 Chapter

INTRODUCTION TO BIOINFORMATICS

INTRODUCTION: AN OVERVIEW

Bioinformatics is an interdisciplinary field that develops methods and software tools for understanding biological data. As an interdisciplinary field of science, bioinformatics combines computer science, statistics, mathematics, and engineering to analyze and interpret biological data. Bioinformatics has been used for in silico analyses of biological queries using mathematical and statistical techniques. Bioinformatics derives knowledge from computer analysis of biological data. These can consist of the information stored in the genetic code, but also experimental results from various sources, patient statistics, and scientific literature. Research in bioinformatics includes method development for storage, retrieval, and analysis of the data. Bioinformatics is a rapidly developing branch of biology and is highly interdisciplinary, using techniques and concepts from informatics, statistics, mathematics, chemistry, biochemistry, physics, and linguistics. It has many practical applications in different areas of biology and medicine.

The National Center for Biotechnology Information (NCBI 2001) defines bioinformatics as: "Bioinformatics is the field of science in which biology, computer science, and information technology merge into a single discipline. There are three important sub- disciplines within bioinformatics: the development of new algorithms and statistics with which to assess relationships among members of large data sets; the analysis and interpretation of various types of data including nucleotide and amino acid sequences, protein domains, and protein structures; and the development and implementation of tools that enable efficient access and management of different types of information.

Even though the three terms: bioinformatics, computational biology and bioinformation infrastructure are often times used interchangeably, broadly, the three may be defined as follows:

- bioinformatics refers to database-like activities, involving persistent sets of data that are maintained in a consistent state over essentially indefinite periods of time;

- computational biology encompasses the use of algorithmic tools to facilitate biological analyses; while
- bioinformation infrastructure comprises the entire collective of information management systems, analysis tools and communication networks supporting biology. Thus, the latter may be viewed as a computational scaffold of the former two.

There are three important sub-disciplines within bioinformatics:

- the development of new algorithms and statistics with which to assess relationships among members of large data sets;
- the analysis and interpretation of various types of data including nucleotide and amino acid sequences, protein domains, and protein structures;
- and the development and implementation of tools that enable efficient access and management of different types of information

DEFINITION OF BIOINFORMATICS - OTHER SOURCES

- Bioinformatics or computational biology is the use of mathematical and informational techniques, including statistics, to solve biological problems, usually by creating or using computer programs, mathematical models or both. One of the main areas of bioinformatics is the data mining and analysis of the data gathered by the various genome projects. Other areas are sequence alignment, protein structure prediction, systems biology, protein-protein interactions and virtual evolution. (source: www.answers.com)
- Bioinformatics is the science of developing computer databases and algorithms for the purpose of speeding up and enhancing biological research. (source: www.whatis.com)
- The application of computer technology to the management of biological information. Specifically, it is the science of developing computer databases and algorithms to facilitate and expedite biological research.(source: Webopedia)
- Bioinformatics is concerned with the creation and development of advanced information and computational technologies to solve problems in biology.

Bioinformatics uses techniques from informatics, statistics, molecular biology and high-performance computing to obtain information about genomic or protein sequence data

A SHORT HISTORY OF BIOINFORMATICS

- 1933 A new technique, electrophoresis, is introduced by Tiselius for separating proteins in solution.
- 1951 Pauling and Corey propose the structure for the alpha-helix and beta-sheet (Proc. Natl. Acad. Sci. USA, 27: 205-211, 1951; Proc. Natl. Acad. Sci. USA, 37: 729- 740, 1951).
- 1953 Watson and Crick propose the double helix model for DNA based on x-ray data obtained by Franklin and Wilkins (Nature, 171: 737-738, 1953).
- 1954 Perutz's group develop heavy atom methods to solve the phase problem in protein crystallography.
- 1955 The sequence of the first protein to be analyzed, bovine insulin, is announced by F. Sanger.
- 1969 The ARPANET is created by linking computers at Stanford and UCLA.
- 1970 The details of the Needleman-Wunsch algorithm for sequence comparison are published.
- 1972 The first recombinant DNA molecule is created by Paul Berg and his group. 1973 The Brookhaven Protein Data Bank is announced (Acta. Cryst. B, 1973, 29: 1746).
- Robert Metcalfe receives his Ph.D. from Harvard University. His thesis describes Ethernet.
- 1974 Vint Cerf and Robert Kahn develop the concept of connecting networks of computers into an "internet" and develop the Transmission Control Protocol (TCP). 1975 Microsoft Corporation is founded by Bill Gates and Paul Allen.
- Two-dimensional electrophoresis, where separation of proteins on SDS polyacrylamide gel is combined with separation according to isoelectric points, is announced by P. H. O'Farrell (J. Biol. Chem., 250: 4007-4021, 1975).
- E. M. Southern published the experimental details for the Southern Blot technique of specific sequences of DNA (J. Mol. Biol., 98: 503-517, 1975).
- 1977 The full description of the Brookhaven PDB (http://www.pdb.bnl.gov) is published (Bernstein, F.C.; Koetzle, T.F.; Williams, G.J.B.; Meyer, E.F.; Brice, M.D.; Rodgers, J.R.; Kennard, O.; Shimanouchi, T.; Tasumi, M.J.; J. Mol. Biol., 1977, 112:, 535).
- Allan Maxam and Walter Gilbert (Harvard) and Frederick Sanger (U.K. Medical Research Council), report methods for sequencing DNA.
- 1980 The first complete gene sequence for an organism (FX174) is published.

- Wuthrich et. al. publish paper detailing the use of multi-dimensional NMR for protein structure determination (Kumar, A.; Ernst, R.R.; Wuthrich, K.; Biochem. Biophys. Res. Comm., 1980, 95:, 1).
- Intelli Genetics, Inc. founded in California. Their primary product is the IntelliGenetics Suite of programs for DNA and protein sequence analysis.
- 1981 The Smith-Waterman algorithm for sequence alignment is published.
- IBM introduces its Personal Computer to the market.
- 1982 Genetics Computer Group (GCG) created as a part of the University of Wisconsin of Wisconsin Biotechnology Center. The company's primary product is The Wisconsin Suite of molecular biology tools.
- 1983 The Compact Disk (CD) is launched.
- 1984 Jon Postel's Domain Name System (DNS) is placed on-line. The Macintosh is announced by Apple Computer.
- 1985 The FASTP algorithm is published.
- The PCR reaction is described by Kary Mullis and co-workers.
- 1986 The term "Genomics" appeared for the first time to describe the scientific discipline of mapping, sequencing, and analyzing genes. The term was coined by Thomas Roderick as a name for the new journal.
- Amoco Technology Corporation acquires IntelliGenetics. NSFnet debuts.
- The SWISS-PROT database is created by the Department of Medical Biochemistry of the University of Geneva and the European Molecular Biology Laboratory (EMBL). 1987 The use of yeast artifical chromosomes (YAC) is described (David T. Burke, et. al., Science, 236: 806-812).
- The physical map of E. coli is published (Y. Kohara, et. al., Cell 51: 319-337).
- 1988 The National Center for Biotechnology Information (NCBI) is established at the National Cancer Institute.
- The Human Genome Initiative is started (Commission on Life Sciences, National Research Council. Mapping and Sequencing the Human Genome, National Academy Press: Washington, D.C.), 1988.
- The FASTA algorithm for sequence comparison is published by Pearson and Lupman. A new program, an Internet computer virus designed by a student, infects 6,000
- 1989 The Genetics Computer Group (GCG) becomes a private company.
- Oxford Molecular Group, Ltd. (OMG) founded in Oxford, UK by Anthony Marchington, David Ricketts, James Hiddleston, Anthony Rees, and W.

Graham Richards. Primary products: Anaconda, Asp, Cameleon and others (molecular modeling, drug design, protein design).

- 1990 The BLAST program (Altschul, et. al.) is implemented.
- Molecular Applications Group is founded in California by Michael Levitt and Chris Lee. Their primary products are Look and SegMod which are used for molecular modeling and protein design.
- Infor Max is founded in Bethesda, MD. The company's products address sequence analysis, database and data management, searching, publication graphics, clone construction, mapping and primer design.
- 1991 The research institute in Geneva (CERN) announces the creation of the protocols which make-up the World Wide Web.
- The creation and use of expressed sequence tags (ESTs) is described (J. Craig Venter, et. al., Science, 252: 1651-1656).
- Incyte Pharmaceuticals, a genomics company headquartered in Palo Alto California, is formed.
- Myriad Genetics, Inc. is founded in Utah. The company's goal is to lead in the discovery of major common human disease genes and their related pathways. The Company has discovered and sequenced, with its academic collaborators, the following major genes: BRCA1, BRCA2, CHD1, MMAC1, MMSC1, MMSC2, CtIP, p16, p19, and MTS2.
- 1992 Human Genome Systems, Gaithersburg Maryland, is formed by William Haseltine.
- The Institute for Genomic Research (TIGR) is established by Craig Venter. Genome Therapeutics announces its incorporation.
- Mel Simon and coworkers announce the use of BACs for cloning. 1993 CuraGen Corporation is formed in New Haven, CT. Affymetrix begins independent operations in Santa Clara, California 1994
- Netscape Comminications Corporation founded and releases Navigator, the commercial version of NCSA's Mozilla.
- Gene Logic is formed in Maryland.
- The PRINTS database of protein motifs is published by Attwood and Beck. Oxford Molecular Group acquires IntelliGenetics.
- 1995 The Haemophilus influenzea genome (1.8 Mb) is sequenced. The Mycoplasma genitalium genome is sequenced.
- 1996 Oxford Molecular Group acquires the MacVector product from Eastman Kodak.

- The genome for Saccharomyces cerevisiae (baker's yeast, 12.1 Mb) is sequenced. The Prosite database is reported by Bairoch, et.al.
- Affymetrix produces the first commercial DNA chips. 1997 The genome for E. coli (4.7 Mbp) is published.
- Oxford Molecular Group acquires the Genetics Computer Group.
- LION bioscience AG founded as an integrated genomics company with strong focus on bioinformatics. The company is built from IP out of the European Molecular Biology Laboratory (EMBL), the European Bioinformatics Institute (EBI), the German Cancer Research Center (DKFZ), and the University of Heidelberg.
- Paradigm Genetics Inc., a company focussed on the application of genomic technologies to nhance worldwide food and fiber production, is founded in Research Triangle Park, NC. deCode genetics publishes a paper that described the location of the FET1 gene, which is responsible for familial essential tremor, on chromosome 13 (Nature Genetics).
- 1998 The genomes for Caenorhabditis elegans and baker's yeast are published. The Swiss Institute of Bioinformatics is established as a non-profit foundation. Craig Venter forms Celera in Rockville, Maryland.
- PE Informatics was formed as a Center of Excellence within PE Biosystems. This center brings together and leverages the complementary expertise of PE Nelson and Molecular Informatics, to further complement the genetic instrumentation expertise of Applied Biosystems.
- Inpharmatica, a new Genomics and Bioinformatics company, is established by University College London, the Wolfson Institute for Biomedical Research, five leading scientists from major British academic centers and Unibio Limited.
- Gene Formatics, a company dedicated to the analysis and prediction of protein structure and function, is formed in San Diego.
- Molecular Simulations Inc. is acquired by Pharmacopeia 1999 deCode genetics maps the gene linked to pre-eclampsia as a locus on chromosome 2p13.
- 2000 The genome for Pseudomonas aeruginosa (6.3 Mbp) is published. The A. thaliana genome (100 Mb) is secquenced.
- The D. melanogaster genome (180Mb) is sequenced. Pharmacopeia acquires Oxford Molecular Group.
- 2001 The human genome (3,000 Mbp) is published. 2002 Chang Gung Genomic Research Center established.

Bioinformatics Center, -Proteomics Center, -Microarray Center

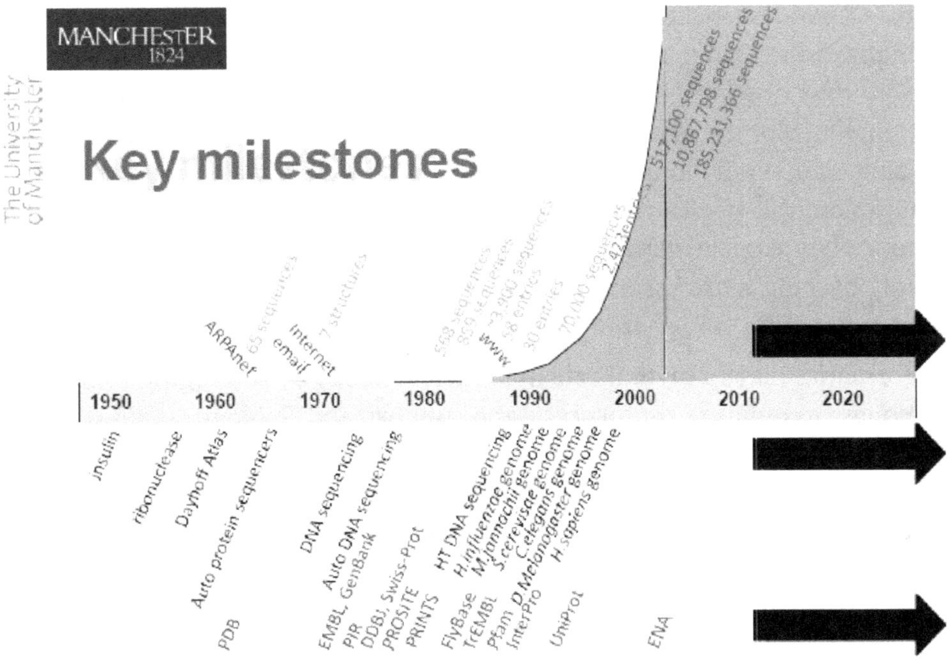

Key Milestones in Bioinformatics

IMPORTANCE IN MODERN BIOLOGY

In recent decades, the intersection of biology and computer science has given rise to a transformative field known as bioinformatics, which has rapidly become a cornerstone of modern biological research. As advancements in molecular biology, genomics, and biotechnology have resulted in an explosion of complex biological data, traditional methods of biological analysis have struggled to keep pace. This is where bioinformatics steps in—bridging the gap between raw biological data and meaningful insights. By harnessing the power of computational algorithms, statistical models, and data visualization techniques, bioinformatics enables researchers to manage, analyze, and interpret vast quantities of data with a level of precision and efficiency that would be impossible with manual methods alone. From sequencing entire genomes to modeling intricate cellular processes, bioinformatics plays an indispensable role in decoding the molecular underpinnings of life. The impact of bioinformatics on modern biology cannot be overstated. It has revolutionized numerous areas of research, enabling scientists to answer fundamental questions about genetics, evolution, and disease. With tools that allow for the rapid analysis of genetic sequences, protein structures, and cellular networks, bioinformatics accelerates discoveries in medicine, agriculture, and environmental sciences. For example, in genomics, bioinformatics facilitates the decoding of complex genomes, aiding in the identification of disease-causing mutations and opening doors to personalized medicine. In drug discovery, bioinformatics tools help researchers identify novel drug targets and design therapeutic compounds more efficiently. Additionally, bioinformatics contributes significantly to ecological studies by analyzing biodiversity and ecosystem dynamics, fostering a deeper understanding of environmental health and sustainability. As we continue to accumulate vast amounts of biological data from high-throughput technologies, bioinformatics will only become more essential in addressing the challenges of modern biology, driving innovation, and shaping the future of life sciences.

Bioinformatics plays a pivotal role in modern biology by leveraging computational techniques to analyze, interpret, and understand biological data. With the explosion of biological data, especially through high-throughput technologies like next-generation sequencing (NGS), bioinformatics has become indispensable in various areas of biological research and applied sciences. Below are some of the key ways in which bioinformatics is important:

Genomics and Genome Sequencing

Genomics, the study of an organism's complete set of DNA, including all of its genes, has been revolutionized by advancements in genome sequencing technologies. Genome sequencing allows researchers to decode the entire genetic blueprint of an organism, providing invaluable insights into its biological functions, evolutionary history, and susceptibility to diseases. Bioinformatics plays a crucial role in genomics by enabling the efficient analysis and interpretation of massive amounts of data generated through high-throughput sequencing methods such as next-generation sequencing (NGS). Through bioinformatic tools, scientists can identify genes, regulatory elements, and non-coding regions that may be crucial for understanding health, disease, and development. Furthermore, bioinformatics allows for the comparison of genomes across species, revealing evolutionary relationships and pinpointing genetic variations associated with specific traits or diseases. This understanding has profound implications for personalized medicine, where genomic data can guide tailored treatments based on an individual's genetic makeup, and for areas such as gene therapy, crop improvement, and the study of microbial communities. In sum, bioinformatics has made it possible to unlock the complex information hidden within genomes, driving progress in a wide array of biological and medical fields.

- **Genome Mapping and Annotation:** Bioinformatics is essential in sequencing genomes of various organisms, including humans, and in identifying genes, regulatory elements, and functional regions. Large-scale sequencing projects, such as the Human Genome Project, would not have been possible without bioinformatics tools for data analysis.

- **Comparative Genomics:** Bioinformatics allows for the comparison of genomic sequences across different species. This helps researchers identify conserved genes, evolutionary relationships, and understand the genetic basis of diseases.

Proteomics and Protein Structure Prediction

Proteomics, the large-scale study of proteins, is an essential branch of modern biology that complements genomics by focusing on the functional elements of the cell. While genomics provides the blueprint of an organism's genetic code, proteomics delves into the dynamic world of proteins—the molecular machines that carry out most biological functions. Proteins are responsible for nearly every process in the cell, from catalyzing metabolic reactions to providing structural support, yet understanding their full range of activities requires more than just sequencing their genes. This is where bioinformatics comes into play. The explosion of proteomic data, generated through techniques like mass spectrometry and protein microarrays, has necessitated advanced computational methods for analyzing and interpreting complex datasets. Bioinformatics tools allow researchers to identify and quantify proteins, track their post-translational modifications, and examine how they interact with one another within cellular networks.Another critical aspect of proteomics is protein structure prediction, a field in which bioinformatics has made groundbreaking advancements. Understanding a protein's three-dimensional structure is crucial for determining its function, stability, and potential as a drug target. However, experimentally determining protein structures is often time-consuming and expensive. Bioinformatics techniques, including computational modeling, molecular dynamics simulations, and structural prediction algorithms like AlphaFold, have significantly accelerated the process of predicting protein structures from amino acid sequences. These tools help scientists infer how proteins fold and how their structures influence biological activity, enabling the design of more effective therapies for diseases such as cancer, Alzheimer's, and cystic fibrosis. In addition to predicting structures, bioinformatics aids in understanding protein-protein interactions, which are critical for cellular communication and disease mechanisms. As our ability to generate proteomic data continues to grow, bioinformatics will remain a central tool in unlocking the complex relationships between genes, proteins, and cellular functions, offering new opportunities for drug discovery, disease understanding, and biotechnology applications.

- **Protein Structure and Function:** Bioinformatics aids in predicting the structure and function of proteins by analyzing amino acid sequences. Tools such as BLAST, PDB (Protein Data Bank), and structural prediction software (e.g., AlphaFold) are essential for understanding protein folding and interactions.

- **Functional Annotation:** Bioinformatic tools can predict the function of proteins by aligning them with known sequences in databases. This is crucial for understanding cellular processes, protein function, and designing drugs.

Systems Biology

Systems biology is an interdisciplinary field that seeks to understand complex biological systems as a whole, rather than focusing on individual components in isolation. It emerged from the recognition that biological processes are not simply the sum of their parts—genes, proteins, and metabolites—but rather emerge from their intricate interactions and feedback loops within a cellular or organismal context. By integrating data from genomics, proteomics, transcriptomics, and other high-throughput technologies, systems biology aims to create comprehensive models that reflect the dynamic and interconnected nature of biological systems. This holistic approach allows scientists to study how cells, tissues, and entire organisms respond to different stimuli, how diseases develop, and how different biological networks interact and influence one another. With the rapid accumulation of omics data, systems biology has become central to understanding the complexity of life itself, moving beyond reductionist methods to address questions at the systems level. Bioinformatics plays an indispensable role in systems biology by providing the computational tools needed to integrate vast amounts of data and generate predictive models of biological networks. These models help researchers understand how molecular components—such as genes, proteins, and metabolites—communicate within cellular pathways to carry out functions like metabolism, cell signaling, and gene expression. Bioinformatics also enables the analysis of high-dimensional data to identify key regulatory nodes in these networks, offering insights into how disruptions in these networks can lead to diseases such as cancer, diabetes, and neurological disorders. One of the most powerful aspects of systems biology is its ability to model disease progression and predict responses to treatments, opening new avenues for personalized medicine. Furthermore, by understanding the interdependencies within biological systems, researchers can identify novel drug targets and design therapies that address the root causes of diseases, rather than just alleviating symptoms. As data generation continues to accelerate, systems biology, fueled by bioinformatics, is poised to be at the forefront of major breakthroughs in biology and medicine, offering a deeper, more integrated understanding of the living world.

- **Pathway Analysis:** Bioinformatics helps in understanding the complex interactions between genes, proteins, and metabolites within biological pathways. By integrating data from genomics, proteomics, and metabolomics, bioinformatics provides insights into cellular functions and disease mechanisms.

- **Network Biology:** It allows the study of molecular networks and systems-level interactions, helping to model diseases such as cancer, diabetes, and neurological disorders.

Drug Discovery and Personalized Medicine

The fields of drug discovery and personalized medicine have been transformed by the integration of bioinformatics, which has significantly enhanced our ability to design targeted therapies and tailor medical treatments to individual patients. Traditionally, drug discovery was a time-consuming and largely trial-and-error process, with researchers screening thousands of chemical compounds to identify potential candidates for treatment. Today, bioinformatics plays a pivotal role in making this process more efficient by using computational tools to predict how different molecules might interact with specific biological targets, such as proteins or enzymes implicated in disease. Through molecular docking, virtual screening, and quantitative structure-activity relationship (QSAR) modeling, bioinformatics can help identify promising drug candidates with high precision, accelerating the path from discovery to clinical application. Furthermore, by analyzing large datasets of genomic, proteomic, and clinical information, bioinformatics enables the identification of novel drug targets, such as specific genetic mutations or disease-associated proteins, that might not have been obvious through conventional methods. Personalized medicine, another critical area in modern healthcare, has also been greatly enhanced by bioinformatics. Personalized or precision medicine seeks to tailor medical treatment to the individual characteristics of each patient, taking into account their unique genetic makeup, lifestyle, and environmental factors. By analyzing a patient's genomic data, bioinformatics tools can identify genetic variations that affect drug metabolism, drug efficacy, and the likelihood of adverse reactions. For example, genetic testing can reveal whether a patient's DNA carries mutations that make them more susceptible to certain diseases, or whether they will respond well to a specific medication. In cancer treatment, bioinformatics allows for the identification of mutations driving tumor growth, enabling the design of targeted therapies that specifically inhibit these mutations while sparing healthy tissue. Moreover, bioinformatics-based algorithms can be used to predict the best treatment options for individual patients, improving outcomes and minimizing side effects. The combination of advanced genomic data analysis and computational modeling has made personalized medicine more accessible, allowing healthcare providers to make more informed, data-driven decisions for their patients. As genomic sequencing continues to decrease in cost and more clinical data becomes available, bioinformatics will be central in making personalized medicine a standard of care, paving the way for more precise and effective treatments in the future.

- **Drug Target Identification:** By analyzing genetic and proteomic data, bioinformatics can identify potential drug targets for diseases. Computational techniques such as molecular docking and virtual screening can help in drug design and optimization.
- **Personalized Medicine:** Bioinformatics enables the analysis of individual genomic data to tailor medical treatments specific to the genetic makeup of patients. This is especially relevant in oncology, where mutations in cancer cells can guide targeted therapies.

Microbial and Metagenomics

Microbial and metagenomics represent two rapidly growing areas of genomics that have been fundamentally transformed by the power of bioinformatics. Microbial genomics focuses on the study of individual microorganisms by sequencing their genomes, revealing critical insights into their biology, evolution, and pathogenic potential. These microorganisms—bacteria, viruses, fungi, and archaea—are ubiquitous in nature and have profound impacts on human health, agriculture, and the environment. By using high-throughput sequencing technologies, bioinformatics tools enable researchers to decode microbial genomes with unprecedented speed and accuracy. This has led to major breakthroughs in understanding the genetics of disease-causing pathogens, including antibiotic-resistant bacteria, as well as beneficial microbes that play essential roles in processes like digestion, nitrogen fixation, and bioremediation. With bioinformatics, scientists can identify key genes involved in microbial virulence, antibiotic resistance, and metabolic pathways, which can inform the development of new diagnostics, treatments, and preventive strategies. Metagenomics, on the other hand, takes a more holistic approach by studying the collective genetic material of microbial communities in a given environment, without the need for culturing individual organisms. Metagenomic sequencing enables the exploration of complex microbial ecosystems found in diverse habitats such as the human gut, soil, oceans, and the atmosphere. Using bioinformatics tools to analyze metagenomic data, researchers can assess the diversity of microbial species, identify novel microorganisms, and investigate their functional roles within their ecosystems. For example, in human health, metagenomics allows for the study of the gut microbiome and its impact on digestion, immunity, and even mental health. In environmental science, metagenomics can be used to monitor biodiversity and understand the roles of microbes in nutrient cycling, pollution degradation, and climate change. Furthermore, metagenomic approaches are instrumental in identifying novel antimicrobial compounds and in discovering new enzymes with industrial applications. Bioinformatics algorithms facilitate the assembly, annotation, and functional analysis of metagenomic sequences, which would be otherwise impossible to manage manually due to the immense scale and complexity of the data. As sequencing technologies continue to advance, the integration of bioinformatics in microbial and metagenomic research will continue to uncover new insights into the microbial world, transforming our understanding of both beneficial and harmful microbes, and opening new frontiers in medicine, agriculture, and environmental conservation.

- **Microbial Genomics:** Bioinformatics is used to sequence and analyze the genomes of microorganisms, leading to advancements in microbiology, including the study of pathogens and antibiotic resistance.
- **Metagenomics:** The study of complex microbial communities, such as those found in the human gut or environmental ecosystems, relies heavily on bioinformatics tools to analyze metagenomic data. This can reveal insights into microbial diversity, ecology, and health-related impacts.

Evolutionary Biology

Evolutionary biology seeks to understand the origins and changes in the diversity of life on Earth, focusing on how organisms evolve over time through processes like natural selection, genetic drift, and gene flow. Central to evolutionary biology is the idea that all species share a common ancestor and have diversified through gradual changes in their genetic makeup. In recent decades, the field has been profoundly reshaped by the advent of high-throughput sequencing technologies and bioinformatics. The ability to sequence entire genomes from both living organisms and ancient fossils has provided an unprecedented wealth of data, allowing scientists to reconstruct evolutionary histories with greater accuracy and detail than ever before. Bioinformatics plays a crucial role in analyzing these vast datasets, enabling researchers to identify genetic variations that contribute to traits such as resistance to diseases, environmental adaptation, and reproductive success. By comparing the genomes of different species, bioinformatics tools can uncover conserved genetic elements that provide insights into evolutionary processes, such as speciation and adaptive radiation, shedding light on the molecular basis of evolution. One of the most powerful applications of bioinformatics in evolutionary biology is phylogenetic analysis, which involves constructing evolutionary trees (phylogenies) to trace the relationships between species based on their genetic similarities and differences. Through computational tools, researchers can create highly detailed phylogenetic trees that not only map the evolutionary history of organisms but also reveal how specific traits evolved and how organisms adapted to their environments over time. Bioinformatics also facilitates comparative genomics, which allows scientists to identify conserved genes and genomic regions across species, providing a deeper understanding of fundamental biological processes like development, aging, and disease. In addition, the field of population genetics—which studies the genetic composition of populations and how it changes over time—relies heavily on bioinformatic methods to analyze large-scale genetic data from different populations, helping to reveal the genetic basis of traits and diseases and the mechanisms driving evolutionary change. As genomic data from an ever-expanding range of species become available, bioinformatics tools will continue to play an essential role in shaping our understanding of evolution, helping us to trace the genetic roots of biodiversity and the molecular underpinnings of adaptation. This comprehensive understanding is vital not only for advancing evolutionary biology but also for fields such as conservation biology, where bioinformatics is used to study biodiversity and the impact of environmental changes on the survival of species.

- **Phylogenetic Analysis:** Bioinformatics allows scientists to construct phylogenetic trees that trace evolutionary relationships among species. These trees are based on genetic data and provide insights into evolutionary processes, speciation, and adaptation.
- **Population Genetics:** Tools like genome-wide association studies (GWAS) help researchers identify genetic variations that contribute to traits or diseases in populations.

Big Data and High-Throughput Biology

In the past few decades, the field of biology has experienced an explosion of data, driven largely by advances in high-throughput technologies that can generate massive volumes of information in a short amount of time. These technologies—such as next-generation sequencing (NGS), microarrays, mass spectrometry, and high-throughput imaging—allow scientists to simultaneously analyze thousands or even millions of biological samples, providing insights into genomics, proteomics, transcriptomics, and other aspects of cellular biology. However, the sheer volume, complexity, and diversity of this data present significant challenges. This is where bioinformatics comes in, serving as the critical bridge between raw biological data and meaningful scientific discoveries. By leveraging advanced computational techniques, machine learning algorithms, and data visualization tools, bioinformatics enables researchers to extract valuable insights from big biological data, turning it into actionable knowledge that can inform drug development, disease diagnosis, and therapeutic strategies. The field of high-throughput biology has revolutionized our ability to explore the biological world at an unprecedented scale. High-throughput sequencing, for instance, allows for the sequencing of entire genomes, transcriptomes, and even epigenomes in a matter of days, enabling comprehensive studies of genetic variation, gene expression, and regulatory networks. Mass spectrometry and proteomics technologies have made it possible to identify and quantify thousands of proteins from complex biological samples, offering insights into cellular function and disease mechanisms. However, the challenge lies not just in generating these enormous datasets, but in managing, analyzing, and interpreting them. Bioinformatics provides the tools to handle this vast amount of data, from storage and processing to statistical analysis and pattern recognition. By integrating data from multiple sources—such as genomics, transcriptomics, and proteomics—bioinformatics enables a more holistic understanding of biological systems, allowing researchers to construct detailed molecular networks, predict cellular responses to various conditions, and model complex biological phenomena. The term big data in the context of high-throughput biology refers not only to the vast quantities of information being generated, but also to the complexity of the data, which often includes diverse data types, irregular formats, and noisy, incomplete information. To tackle these challenges, bioinformatics employs sophisticated algorithms and machine learning techniques that can identify patterns, correlations, and causal relationships within the data, even in the face of inherent biological variability. Furthermore, bioinformatics tools allow for the integration of different layers of biological information (e.g., genetic, proteomic, and metabolomic), facilitating a systems biology approach to understanding health and disease. With advancements in cloud computing and data storage, researchers can now share and collaborate on large-scale biological datasets globally, accelerating scientific discovery. In fields such as precision medicine, cancer research, and microbiome studies, big data is increasingly being harnessed to uncover new biomarkers, identify therapeutic targets, and predict patient outcomes with greater accuracy. As data generation continues to accelerate, bioinformatics will remain an essential tool for managing and extracting meaningful insights from big biological data, driving forward both basic and applied biological research and transforming our understanding of life at the molecular level.

- **Handling Large Datasets:** Modern biology generates enormous amounts of data, from sequencing data to imaging data. Bioinformatics is crucial for storing, managing, and analyzing these large datasets using tools for data mining, machine learning, and statistical analysis.
- **Data Integration:** Bioinformatics enables the integration of various types of biological data (genomic, proteomic, transcriptomic, etc.) to provide a more comprehensive understanding of biological processes.

Agricultural and Environmental Applications

Bioinformatics has become an indispensable tool in both agriculture and environmental science, where it is helping to address some of the most pressing challenges facing our planet today—ranging from food security and crop improvement to biodiversity conservation and climate change. In agriculture, bioinformatics is transforming the way scientists develop new crop varieties, improve yields, and protect crops from pests and diseases. By applying genomic tools to analyze plant and animal genomes, bioinformatics enables the identification of key genes responsible for desirable traits, such as drought tolerance, disease resistance, and nutritional quality. This knowledge is then used to guide traditional breeding programs, as well as in the development of genetically modified (GM) crops that can withstand environmental stressors, grow in suboptimal soils, or offer enhanced nutritional value. Moreover, bioinformatics tools allow for the identification of genetic markers linked to these traits, enabling marker-assisted selection (MAS), a faster and more precise method of crop improvement that shortens the breeding cycle and enhances the efficiency of crop development.In addition to crop improvement, bioinformatics is crucial for managing agricultural biodiversity and protecting crops from emerging pathogens. The sequencing of microbial genomes in agricultural ecosystems, including soil microbes and plant pathogens, has revolutionized our understanding of plant-microbe interactions. Bioinformatic tools can identify microbial communities that enhance soil health, fix nitrogen, or suppress harmful pathogens, contributing to more sustainable farming practices. Additionally, bioinformatics allows for the identification of genetic variants in pathogens that confer resistance to pesticides or herbicides, helping to design more effective management strategies. Understanding the genetic diversity of pests and pathogens is also critical for combating emerging diseases, such as plant viral epidemics, that threaten food security worldwide. Bioinformatics is equally important in environmental science, particularly in the study of ecosystems, biodiversity, and the impact of environmental changes on species and habitats. In conservation biology, the ability to sequence the genomes of endangered species or monitor the genetic diversity of populations is critical for making informed decisions about conservation strategies. With the help of bioinformatics, researchers can track the genetic health of populations, understand how species adapt to environmental pressures, and identify genetic markers that could be used in conservation programs to prevent inbreeding and loss of genetic diversity. Additionally, bioinformatics is essential in the study of environmental microbiomes—the diverse microbial communities present in air, water, and soil. Through metagenomic sequencing and analysis, scientists can gain insights into microbial diversity, their role in nutrient cycling, and their response to environmental changes, such as pollution or climate shifts. In the context of climate change, bioinformatics is being used to study the genetic basis of adaptation in both plants and animals, allowing researchers to predict how species may evolve in response to changing temperatures, drought, and other climate stressors. This information is crucial for developing resilient agricultural systems and ensuring that ecosystems can withstand environmental disturbances. Moreover, bioinformatics-driven models of ecosystem dynamics help predict the potential impacts of environmental changes on biodiversity and the provision of ecosystem services, such as pollination, water purification, and carbon sequestration. Finally, sustainable agriculture and environmental management benefit from bioinformatics by improving our ability to monitor and manage the impact of farming practices on ecosystems. Through the analysis of environmental DNA (eDNA), researchers can assess biodiversity in natural and agricultural environments, detect the presence of invasive species, and monitor ecosystems for early signs of ecological imbalance. These insights help guide conservation efforts and promote practices that protect natural resources while maintaining agricultural productivity. Bioinformatics is revolutionizing both agriculture and environmental science, providing the tools needed to create more resilient, sustainable systems that can meet the challenges posed by climate change, population growth, and environmental degradation. By enabling precise genetic analysis, improving crop yields, enhancing biodiversity conservation, and promoting ecosystem sustainability, bioinformatics plays a pivotal role in ensuring the health of our planet and its resources for future generations.

- **Crop Improvement**: Bioinformatics is used in agricultural biotechnology for the identification of genes associated with desirable traits like drought resistance, pest resistance, and higher yield. It aids in marker-assisted breeding and genetic modification of crops.
- **Environmental Biology:** Bioinformatics tools are used to monitor biodiversity, study ecosystem dynamics, and assess the impacts of environmental changes on organisms, particularly through the analysis of environmental DNA (eDNA).

Data Sharing and Collaboration

In the age of big data and high-throughput biology, data sharing and collaboration have become critical components in advancing scientific discovery, particularly in the life sciences. The immense volume of biological data generated by genomic sequencing, proteomics, metabolomics, and other high-throughput technologies presents both a challenge and an opportunity. On one hand, the sheer scale of this data makes it difficult for any single research group or institution to handle all aspects of its collection, analysis, and interpretation. On the other hand, when shared and collaboratively analyzed, these large datasets can reveal insights that no individual team could uncover on its own. Bioinformatics has facilitated this paradigm shift by providing tools and platforms that enable seamless sharing, integration, and analysis of diverse biological data from multiple sources and across various disciplines. Public biological databases are one of the key innovations that have fueled collaboration and data sharing in modern biology. Repositories like GenBank, Ensembl, Protein Data Bank (PDB), and the European Bioinformatics Institute (EBI) allow researchers worldwide to deposit, access, and compare genetic sequences, protein structures, and other types of biological data. These publicly available databases serve as essential resources for researchers, enabling them to share their findings and build on the work of others. For instance, genomic sequences of model organisms, pathogens, or human populations are now widely accessible, promoting global collaboration on issues such as disease research, drug development, and environmental sustainability. This open-access model fosters a more transparent scientific process, where results can be validated and replicated, and new hypotheses can be tested more quickly. In addition to databases, collaborative platforms have become integral to modern research. Cloud-based computing solutions, such as Google Cloud, Amazon Web Services (AWS), and platforms like Galaxy, allow scientists to store, process, and analyze large biological datasets in a scalable and flexible manner. These platforms also facilitate real-time collaboration, enabling multidisciplinary teams to work together across geographic locations and institutions. Whether it is analyzing genomic data for disease-associated mutations, modeling protein interactions, or predicting the effects of environmental changes on ecosystems, these collaborative environments allow researchers to pool their expertise and resources, speeding up the research process and increasing the overall impact of their work. Collaborative networks are especially important in large-scale research initiatives. Projects like the Human Genome Project, Cancer Genome Atlas, and Earth BioGenome Project rely on the collective efforts of researchers from various countries, institutions, and scientific disciplines. By sharing data and resources across borders, these initiatives are able to tackle complex problems that require a coordinated global effort. Bioinformatics facilitates such collaborations by providing common tools, languages, and frameworks for data analysis, ensuring that results can be integrated and compared, regardless of where the data originated. The ability to work together in this way has led to breakthroughs in understanding the genetic basis of diseases, discovering new drug targets, and mapping biodiversity across the globe. Furthermore, interdisciplinary collaboration—involving biologists, bioinformaticians, data scientists, engineers, and clinicians—has become increasingly common and necessary. Biological data analysis now often requires expertise not just in biology, but also in areas like statistics, machine learning, and data visualization. By bringing together experts from multiple fields, scientists are able to develop more sophisticated models, predictive algorithms, and tools that can provide deeper insights into biological processes and health-related issues. For example, in personalized medicine, bioinformatics professionals collaborate with clinicians and geneticists to develop predictive models for patient outcomes based on their genetic makeup, while also integrating environmental and lifestyle data to improve treatment strategies. In sum, data sharing and collaboration are foundational to modern scientific progress, and bioinformatics plays a crucial role in facilitating both. By enabling access to large-scale biological datasets, providing computational tools for collaborative analysis, and fostering interdisciplinary networks, bioinformatics helps break down the silos that have traditionally existed in scientific research. The result is a more connected, transparent, and efficient scientific community, where discoveries are accelerated, validated more rigorously, and applied to real-world problems more quickly. In the context of global health challenges, climate change, and biodiversity conservation, the collaborative use of bioinformatics and shared data has never been more essential to driving forward our understanding and solutions to the complex issues facing our world.

- **Public Databases:** Bioinformatics promotes open access to biological data through public repositories like GenBank, Ensembl, and others. This fosters collaboration across the global scientific community, accelerating discoveries and innovation.
- **Visualization Tools:** Data visualization is a crucial aspect of bioinformatics, helping researchers present complex biological information in intuitive ways, such as through interactive genomics browsers or 3D protein structure viewers.

Bioinformatics has become an essential field that bridges the gap between biology and computational sciences. It is foundational to almost every area of modern biology, enabling researchers to explore biological systems in unprecedented detail. As technology continues to evolve, bioinformatics will remain at the forefront, driving innovations in medicine, agriculture, environmental science, and beyond.

INTERDISCIPLINARY NATURE OF BIOINFORMATICS

Bioinformatics stands at the intersection of multiple scientific disciplines, drawing upon knowledge and techniques from biology, computer science, mathematics, physics, engineering, and statistics to solve complex biological problems. This interdisciplinary nature is one of the key reasons why bioinformatics has become such a powerful tool in modern biology. At its core, bioinformatics is about applying computational techniques to analyze and interpret large volumes of biological data, but to do this effectively, bioinformaticians must not only understand the underlying biological concepts but also have a deep knowledge of algorithm design, data structures, and statistical modeling. This unique combination of biology and computer science is what makes bioinformatics so versatile and essential in driving advancements in fields like genomics, proteomics, personalized medicine, and environmental science. From the biological side, bioinformaticians must have a solid foundation in molecular biology, genetics, and biochemistry. This is crucial for understanding the complex data they are working with—whether it's genomic sequences, protein structures, or metabolic networks. Understanding the biology behind the data allows bioinformaticians to ask the right questions, identify meaningful patterns, and interpret results accurately. For example, in genomic research, bioinformaticians need to understand the intricacies of gene expression, regulatory pathways, and genetic mutations to properly analyze sequencing data and draw biologically relevant conclusions. In proteomics, they must understand protein folding, post-translational modifications, and the functional roles of proteins within a cell. Without this biological knowledge, even the most sophisticated computational models would lack context and the ability to address the real-world challenges of biology. On the computational side, bioinformaticians need expertise in computer science and software engineering. Developing and optimizing algorithms to process and analyze large biological datasets requires an understanding of computational complexity, data structures, and software design. Bioinformatics also relies heavily on database management, as vast amounts of biological data—such as gene sequences, protein structures, and clinical data—must be stored, retrieved, and manipulated efficiently. Furthermore, bioinformaticians employ machine learning and artificial intelligence techniques to identify patterns in large datasets, making predictions about gene function, disease mechanisms, or drug efficacy. This requires knowledge of statistical modeling, data mining, and algorithmic optimization. The integration of these computational approaches with biological data allows bioinformaticians to create predictive models, analyze genetic variation, and understand complex biological systems at a systems level.

The interdisciplinary nature of bioinformatics also extends to its application in clinical and medical settings. For personalized medicine, bioinformaticians work closely with clinicians, geneticists, and pharmacologists to analyze patient data, genetic profiles, and treatment responses to develop customized therapeutic strategies. In cancer research, bioinformaticians collaborate with oncologists to analyze tumor genomics and identify biomarkers that predict how patients will respond to specific treatments. In drug discovery, bioinformaticians team up with chemists and pharmacologists to analyze protein targets and design small molecules or biologics that can modulate their activity. This integration of bioinformatics with other fields is essential for translating basic biological insights into tangible medical applications that can improve patient outcomes. Moreover, bioinformaticians contribute to epidemiological studies, working with public health experts to track disease outbreaks, identify genetic risk factors for diseases, and predict future health trends. In addition, bioinformatics also draws heavily on mathematics and statistics, especially when analyzing high-dimensional data like genomic sequences, gene expression profiles, or population-level genetic studies. Statistical models are used to assess the significance of results, identify trends, and make predictions based on large and often noisy datasets. Machine learning algorithms and artificial intelligence are increasingly used to automate the analysis of biological data, uncover hidden patterns, and make accurate predictions about disease risk, drug response, or protein-protein interactions. For example, clustering algorithms are used to categorize gene expression patterns, while machine learning models can predict protein folding or the potential for drug molecules to interact with specific targets. This heavy reliance on quantitative analysis requires a strong background in mathematics, statistics, and computational modeling, making bioinformatics inherently interdisciplinary. Finally, engineering and physics also contribute to bioinformatics, particularly in areas like structural bioinformatics and systems biology. In structural bioinformatics, physicists and engineers apply principles of physics to understand protein folding, molecular dynamics, and the forces that govern biomolecular interactions. These fields also contribute to the development of computational tools for simulating biological processes and predicting the behavior of biomolecules. In systems biology, the mathematical and computational modeling of complex biological networks—such as gene regulatory networks, metabolic pathways, and protein interaction networks—requires expertise in systems theory, dynamical systems, and control theory, all of which are drawn from engineering and physics. The interdisciplinary approach is essential for tackling the complex challenges that modern biology presents. As biological questions grow in scope and complexity—especially with the advent of big data and high-throughput technologies—the ability to collaborate across fields is more important than ever. The challenges of data integration, network modeling, predictive analytics, and system-level understanding require the combined expertise of biologists, computer scientists, mathematicians, engineers, and clinicians. Bioinformatics, as a field, thrives precisely because it integrates these diverse perspectives, enabling groundbreaking discoveries across biology, medicine, and environmental science. By bringing together expertise from multiple disciplines, bioinformatics is not only advancing our understanding of life itself but also driving innovations in health, sustainability, and biotechnology.

Bioinformatics is inherently interdisciplinary, as it sits at the convergence of multiple scientific fields, combining knowledge and techniques from **biology**, **computer science**, **mathematics**, **statistics**, **engineering**, and **physics**. This interdisciplinary nature enables bioinformatics to address some of the most complex challenges in modern biology, from understanding the molecular basis of disease to creating new drugs and improving agricultural practices. The need to draw from such diverse disciplines is a direct consequence of the massive amounts of data being generated in modern biological research and the sophisticated analytical approaches required to interpret and make sense of it.

Biology: The Foundation of Bioinformatics

At the core of bioinformatics lies **biology**, as it provides the context and content for the data that bioinformaticians analyze. In order to effectively process, interpret, and draw meaningful conclusions from biological data, bioinformaticians must have a solid understanding of biological concepts such as **genetics, molecular biology**, **biochemistry**, **genomics**, and **proteomics**. Understanding the structure and function of biological molecules, including genes, proteins, and metabolites, is fundamental for interpreting sequencing data and making predictions about gene function, metabolic pathways, or protein interactions. For example, bioinformaticians working with **genomic data** need to understand the principles of **gene expression, mutations**, and **regulation** to accurately interpret sequencing results. Similarly, in **proteomics**, understanding the folding, structure, and function of proteins is essential for predicting how changes in the genome can impact cellular function. Without biological knowledge, computational tools would simply generate raw data that might lack biological relevance. Thus, bioinformaticians must be able to bridge the gap between the complexity of biological systems and the power of computational models to analyze and predict their behavior.

Computer Science: Algorithm Development and Data Management

Computer science is the backbone of bioinformatics, as it provides the computational infrastructure required to analyze vast amounts of biological data. The role of computer science in bioinformatics is vast, and it encompasses areas such as **algorithm design**, **data structures**, **database management**, and **software engineering**.

- **Algorithm Development**: Bioinformatics heavily relies on the creation of efficient algorithms that can process large-scale biological data. For example, algorithms are needed for sequence alignment (e.g., **BLAST**), gene prediction, and identifying genetic variants in sequencing data. These algorithms are designed to handle the vast complexity of biological data, often in the form of long DNA sequences or protein structures, and to do so in a time-efficient manner. Designing algorithms that can accurately map, compare, or assemble sequences is a central aspect of bioinformatics.
- **Data Structures and Database Management**: Biological data often comes in a variety of formats, from genomic sequences to protein structures. To manage this data effectively, bioinformatics makes use of **data structures** like trees, graphs, and matrices, and develops databases that store, retrieve, and organize biological information. Public databases like **GenBank**, **Ensembl**, and the **Protein Data Bank** (PDB) house genetic and protein information that bioinformaticians access to conduct their analyses. The

development and optimization of databases and data storage systems are critical to ensuring that researchers can quickly and easily access large-scale data.

- **Software Engineering:** The design and implementation of computational tools are essential for bioinformaticians. Developing user-friendly software packages, data visualization tools, and web interfaces allows biologists and researchers from various fields to access and utilize bioinformatics resources without requiring deep programming knowledge.

Mathematics and Statistics: Model Building and Data Analysis

Mathematics and statistics provide the quantitative tools that are essential for analyzing complex biological datasets. High-throughput technologies, such as next-generation sequencing (NGS) or mass spectrometry, generate large volumes of data that require sophisticated statistical techniques to analyze and interpret.

- **Statistical Models:** Bioinformaticians use **statistical models** to quantify uncertainty, test hypotheses, and determine the significance of biological findings. For instance, in **genome-wide association studies** (GWAS), statistical methods are applied to identify genetic variants associated with specific traits or diseases by comparing genetic data across different populations. Likewise, in **differential gene expression analysis**, statistical tests are used to determine which genes are significantly up- or down-regulated under certain conditions.
- **Data Mining and Machine Learning:** With the advent of big data in biology, bioinformatics has embraced techniques from **machine learning** and **artificial intelligence** (AI). These approaches allow for the identification of hidden patterns and relationships within biological datasets, such as predicting protein structures, inferring gene regulatory networks, or classifying cells based on gene expression. Algorithms such as **random forests**, **support vector machines**, and **deep learning** models can be trained on large datasets to make accurate predictions about biological processes, drug responses, or disease susceptibility.
- **Computational Modeling:** In systems biology, bioinformaticians use **mathematical models** to represent biological systems, such as metabolic networks, signaling pathways, or cellular processes. These models are often dynamic and require knowledge of differential equations and systems theory to simulate biological behavior under different conditions, providing insights into the interactions and feedback loops that govern cellular life.

Physics and Engineering

Structural Bioinformatics and Simulations: Physics and **engineering** also contribute significantly to bioinformatics, especially in areas like **structural bioinformatics** and **biophysical simulations.** The principles of physics help bioinformaticians understand the forces that shape molecular structures and the dynamic behaviors of biomolecules.

- **Molecular Modeling and Simulations**: Bioinformaticians often rely on **molecular dynamics simulations** to predict how proteins fold, how molecules interact, and how biological processes unfold over time. These simulations require knowledge of physics, particularly thermodynamics and kinetics, to model molecular interactions at the atomic level. For example, **protein-protein interactions** (PPIs) can be modeled using physics-based simulations to predict how two proteins might bind and function together.
- **Structural Bioinformatics:** In structural bioinformatics, bioinformaticians use computational tools to predict or analyze the three-dimensional structures of biological molecules, such as proteins and nucleic acids. Understanding the 3D conformation of a molecule is essential for determining its function and potential as a drug target. Physics-based principles are used to model atomic interactions and predict stable protein conformations from sequence data. This information is critical for drug discovery, as it allows scientists to design molecules that specifically target disease-related proteins.
- **Engineering of Bioinformatics Tools**: Engineering knowledge is also crucial in the development of bioinformatics tools that can handle the increasing complexity of biological data. This includes the development of hardware for faster data processing (e.g., high-performance computing) and the creation of software that can scale with increasing data volumes. Additionally, bioinformatics pipelines often integrate with cloud computing platforms and require expertise in **distributed computing** and **parallel processing** to handle large datasets efficiently.

The Collaborative Nature of Bioinformatics

Given its interdisciplinary nature, bioinformatics thrives in a **collaborative environment**. Researchers often work in teams that combine domain-specific knowledge from various fields. For example, in **personalized medicine**, bioinformaticians collaborate with clinicians, geneticists, and pharmacologists to analyze patient data and predict treatment outcomes. In **systems biology**, bioinformaticians team up with biologists and engineers to model complex biological networks. Furthermore, the sharing of data and tools through **public**

repositories and **collaborative platforms** accelerates the pace of discovery and promotes cross-disciplinary learning.

The interdisciplinary nature of bioinformatics is what makes it such a powerful and dynamic field. By combining the expertise of biologists, computer scientists, mathematicians, physicists, and engineers, bioinformatics is able to address some of the most challenging questions in modern biology. Whether it's analyzing genetic variation, predicting protein structures, developing new drug therapies, or modeling complex biological systems, bioinformatics relies on the integration of diverse scientific disciplines to provide deeper insights into the molecular mechanisms of life. As biological data continues to grow in volume and complexity, the collaborative, interdisciplinary approach that defines bioinformatics will remain essential for advancing our understanding of biology and improving human health and the environment.

2 Chapter BIOLOGICAL DATA

INTRODUCTION

Biological data in bioinformatics encompasses a vast array of information derived from living organisms, which is essential for understanding the molecular mechanisms that govern life processes. This data can be broadly categorized into genomic, transcriptomic, proteomic, and metabolomic datasets, each representing different levels of biological complexity. At the heart of bioinformatics is the integration and analysis of these large and often heterogeneous datasets to uncover insights about biological functions, diseases, and evolution. With the advent of high-throughput technologies, such as next-generation sequencing (NGS), microarrays, and mass spectrometry, bioinformatics has become a cornerstone of modern biological research. These tools allow for the generation of immense volumes of data, which require sophisticated computational techniques for storage, processing, analysis, and visualization. The role of bioinformatics has expanded dramatically, not only in fundamental biology but also in applied fields such as personalized medicine, drug discovery, agriculture, and environmental science. By enabling the analysis of entire genomes, transcriptomes, proteomes, and more, bioinformatics provides a deeper understanding of the molecular basis of diseases, allowing for the identification of biomarkers, potential drug targets, and personalized treatment options. Furthermore, bioinformatics plays a crucial role in evolutionary studies by helping scientists track genetic changes over time and across different species, offering insights into the history of life on Earth. As biological data continues to grow in volume and complexity, the field of bioinformatics is constantly evolving, integrating new computational methods such as machine learning and artificial intelligence to handle the vast datasets and uncover patterns that were previously unimaginable. In this context, bioinformatics serves as the bridge between biology and technology, making it a key discipline in the future of science and healthcare.

Biological data refers to the information obtained from biological systems, ranging from molecules such as DNA, RNA, and proteins to more complex data sets like whole genomes or ecosystems. These data types are used in bioinformatics to better understand biological processes, identify disease mechanisms, and explore the evolution of life. Biological data is often large, complex, and diverse, and it comes in many forms, each of which can be analyzed for specific insights into the molecular and cellular functioning of organisms. Below is a detailed description of the key types of biological data typically encountered in bioinformatics:

Genomic Data

Genomic data refers to the complete set of DNA sequences that make up an organism's genome. This includes not just the coding regions (genes) but also the non-coding regions of the DNA that play critical roles in gene regulation and other cellular processes. Genomic data can be obtained using high-throughput sequencing technologies such as next-generation sequencing (NGS), which allows researchers to sequence entire genomes quickly and affordably.

Key elements of genomic data include:

- **DNA Sequences:** The raw sequence data generated from the genome, usually presented in FASTA format.
- **Genomic Annotations:** Information about the locations and functions of genes, exons, introns, promoters, enhancers, and other regulatory elements.
- **Variants:** Changes in the DNA sequence compared to a reference genome, such as single nucleotide polymorphisms (SNPs), insertions, deletions (INDELs), and structural variations.
- **Epigenetic Modifications:** Modifications to DNA or histone proteins (e.g., methylation) that regulate gene expression without changing the underlying DNA sequence.

Transcriptomic Data

Transcriptomic data refers to the set of all RNA molecules expressed in a cell or tissue at a specific point in time. This includes messenger RNA (mRNA), which is transcribed from DNA and translated into proteins, as well as non-coding RNAs like ribosomal RNA (rRNA) and microRNAs (miRNAs). Transcriptomic data is crucial for understanding gene expression patterns, the regulation of genes, and how these patterns change in response to different conditions, such as disease or environmental stress.

Key elements of transcriptomic data include:

- **mRNA Sequences:** Information on the types and quantities of mRNA transcripts present in a sample, obtained through RNA sequencing (RNA-Seq) or microarray technologies.

- **Gene Expression Levels:** The relative abundance of mRNA for each gene, which helps determine which genes are active or silenced under various conditions.
- **Alternative Splicing:** Variants of RNA transcripts produced from the same gene due to differential splicing of the precursor mRNA.
- **Non-Coding RNAs:** Includes a variety of regulatory RNAs, such as miRNAs, long non-coding RNAs (lncRNAs), and small interfering RNAs (siRNAs), which have critical roles in gene regulation and cellular processes.

Proteomic Data

Proteomic data involves the study of proteins, which are the molecular machines that carry out most of the functions in a cell. Proteomics focuses on identifying proteins, determining their structures, understanding their functions, and studying how proteins interact with each other and with other cellular components.

Key elements of proteomic data include:

- **Protein Sequences:** The amino acid sequence of proteins, typically determined through mass spectrometry or protein databases.
- **Protein Expression Levels:** The abundance of proteins in a given sample, which can be quantified using techniques like mass spectrometry or two-dimensional gel electrophoresis.
- **Post-Translational Modifications (PTMs):** Modifications to proteins after they are synthesized, such as phosphorylation, acetylation, ubiquitination, or glycosylation, which can affect their activity, stability, and function.
- **Protein-Protein Interactions:** Data on how proteins interact with each other to form complexes, pathways, and networks essential for cellular functions.

Metabolomic Data

Metabolomics is the study of metabolites, which are small molecules involved in cellular metabolism. Metabolomics provides a snapshot of the metabolic state of a cell or organism at any given time. It can be used to identify biomarkers for diseases, understand metabolic disorders, or explore how environmental factors influence metabolic pathways.

Key elements of metabolomic data include:

- **Metabolite Identification:** The process of identifying small molecules (metabolites) within biological samples, often using mass spectrometry or nuclear magnetic resonance (NMR) spectroscopy.

- **Metabolic Pathways:** The networks of biochemical reactions that govern metabolism in an organism. Metabolomics can help identify which pathways are upregulated or downregulated in response to disease or treatment.
- **Metabolite Quantification:** The measurement of metabolite concentrations in a biological sample to understand metabolic flux and the balance of metabolic processes.

Other Biological Data Types

In addition to genomic, transcriptomic, proteomic, and metabolomic data, there are several other important types of biological data that contribute to a more comprehensive understanding of living systems. One such category is microbiome data, which refers to the genetic material of microbial communities inhabiting various parts of the body, such as the gut, skin, and oral cavity. These microbial communities play a crucial role in health, influencing digestion, immune function, and even mental health. Microbiome data is typically collected through high-throughput sequencing techniques, such as 16S rRNA gene sequencing or shotgun metagenomics, which allow for the identification and quantification of microbial species. Structural data is another essential data type, focusing on the three-dimensional structures of biological macromolecules like proteins, nucleic acids, and complex molecular assemblies. Understanding the structure of these molecules is key to deciphering their function, interactions, and potential roles in diseases. Structural data is often derived from techniques such as X-ray crystallography, nuclear magnetic resonance (NMR), and cryo-electron microscopy (cryo-EM). Clinical data, which includes patient health records, medical histories, diagnoses, treatments, and response outcomes, is increasingly integrated with molecular data to develop personalized medicine strategies and predict disease progression. Collectively, these diverse data types complement genomic and molecular insights, offering a more holistic view of biological systems and disease mechanisms.

- **Microbiome Data:** The study of microbial communities that live in and on an organism. These communities, especially in the gut, play crucial roles in digestion, immunity, and health. Microbiome data is usually collected via DNA sequencing techniques such as 16S rRNA sequencing or shotgun metagenomics.
- **Structural Data:** The 3D structures of biological macromolecules (proteins, nucleic acids) are essential for understanding their function. Structural data is typically derived from techniques such as X-ray crystallography, nuclear magnetic resonance (NMR) spectroscopy, or cryo-electron microscopy (cryo-EM).
- **Clinical Data:** In translational bioinformatics, clinical data—such as patient medical histories, disease diagnoses, treatment responses, and genetic

information—are integrated with molecular biological data to personalize medicine and predict health outcomes.

Data Integration and Analysis

The key challenge in bioinformatics is not only collecting these diverse types of biological data but also integrating them into cohesive models. The integration of genomic, transcriptomic, proteomic, and metabolomic data provides a more holistic view of biological systems and diseases. This requires sophisticated computational tools, algorithms, and databases for:

- **Data Preprocessing:** Cleaning, normalizing, and transforming raw data into usable formats.
- **Data Mining and Pattern Recognition:** Applying statistical methods, machine learning, or artificial intelligence to detect patterns in complex biological datasets.
- **Network and Pathway Analysis:** Understanding how genes, proteins, and metabolites are interconnected in biological pathways, and how these networks change under different conditions.
- **Predictive Modeling:** Using data-driven models to predict biological behaviors, such as how a gene mutation might lead to disease or how a drug will interact with a protein.

Biological data is an essential foundation of bioinformatics and modern biology. Each type of biological data—genomic, transcriptomic, proteomic, metabolomic—provides different, yet complementary insights into the structure, function, and regulation of living systems. Together, these data types enable researchers to address fundamental questions in biology, understand the molecular basis of diseases, and develop more effective strategies for personalized medicine, drug discovery, and biotechnology applications. The continuous advancement in technologies for data generation and computational tools for analysis is pushing the boundaries of what we can discover about life at the molecular level, providing exciting opportunities for innovation across multiple scientific disciplines.

TYPES OF BIOLOGICAL DATA: DNA, RNA, PROTEIN

Biological data encompasses a wide range of molecular information that is crucial for understanding the underlying processes of life. Among the most fundamental types of biological data are DNA, RNA, and protein data, each representing a different level of biological organization. DNA (Deoxyribonucleic Acid) serves as the blueprint for life, carrying the genetic information that dictates the structure and function of an organism. The sequence of nucleotides in DNA encodes the instructions for building proteins, which are the molecular machines that perform most cellular functions. RNA (Ribonucleic Acid) plays a critical role in the process of gene expression. Messenger RNA (mRNA) acts as the intermediary, carrying genetic instructions from DNA to ribosomes, where proteins are synthesized. Additionally, other forms of RNA, such as transfer RNA (tRNA) and ribosomal RNA (rRNA), are involved in translation and the formation of protein complexes. Proteins, the final products of gene expression, are complex molecules made up of amino acids and are essential for virtually every function in the cell, from catalyzing biochemical reactions to supporting cell structure and mediating cellular communication. The interaction between DNA, RNA, and proteins forms the core of molecular biology, and the study of these data types—through technologies such as DNA sequencing, RNA-Seq, and mass spectrometry—provides valuable insights into genetics, disease mechanisms, and the molecular basis of life itself.

Biological data can be broadly categorized into several key types, with **DNA, RNA,** and **protein** data being fundamental to understanding cellular and molecular biology. These data types represent different levels of biological information and are central to the processes of gene expression and regulation, as well as cellular function. Below is a detailed description of each:

DNA Data

DNA (Deoxyribonucleic Acid) is the molecular blueprint of life. It carries the genetic instructions necessary for the growth, development, functioning, and reproduction of all living organisms. DNA is composed of two long strands that form a double helix structure, and it is made up of four nucleotide bases: adenine (A), thymine (T), cytosine (C), and guanine (G). The sequence of these nucleotides encodes genetic information.

Key aspects of DNA data include:

- **DNA Sequences**: The most fundamental form of DNA data is the sequence of nucleotides that make up a specific segment of DNA. This can be obtained through various sequencing technologies, such as **Sanger sequencing** or **Next-Generation Sequencing (NGS)**. DNA sequencing provides the exact order of bases in a segment of DNA and is foundational for understanding genes and their mutations.
- **Genomic Structure**: DNA data also includes information about the structural features of genomes. This involves **genes**, **introns**, **exons**, **promoters**, **enhancers**, and **repetitive elements**. Genomic databases, such as **GenBank** or **Ensembl**, store annotated DNA sequences, helping researchers study the organization and function of genomes.
- **Genetic Variants**: DNA data can reveal genetic variations that can influence traits or contribute to diseases. This includes **single nucleotide**

polymorphisms (SNPs), insertions and deletions (INDELs), and **structural variations**, which are critical for understanding genetic predispositions to diseases, drug responses, and evolutionary processes.

- **Epigenetics:** DNA data also encompasses epigenetic modifications, such as **DNA methylation** and **histone modifications**, which do not change the underlying genetic sequence but regulate gene expression. These modifications play important roles in development, disease, and environmental responses.

RNA Data

RNA (Ribonucleic Acid) acts as an intermediary between DNA and protein synthesis. RNA is transcribed from DNA and carries the genetic information from the nucleus to the ribosomes, where proteins are synthesized. However, RNA is not just a messenger—there are various forms of RNA that play distinct roles in regulating and executing biological processes.

Key aspects of RNA data include:

- **mRNA (Messenger RNA):** mRNA is a type of RNA that is transcribed from DNA and serves as a template for protein synthesis in a process known as **translation.** The abundance of mRNA for specific genes can be used to measure gene expression levels. mRNA data is often obtained through **RNA sequencing (RNA-Seq)**, which provides information about the presence, quantity, and variation of mRNA in a sample.
- **Alternative Splicing:** RNA data also includes information on **alternative splicing**, a process where different exons of a gene are joined together to produce multiple mRNA variants from a single gene. This is crucial for understanding how a single gene can give rise to multiple proteins with different functions.
- **Non-Coding RNAs:** While mRNA is the most well-known form of RNA, a significant portion of the genome transcribes non-coding RNAs, which do not encode proteins but serve regulatory functions. This includes:
- **rRNA (Ribosomal RNA):** rRNA is a structural component of ribosomes, the cellular machinery responsible for protein synthesis.
- **tRNA (Transfer RNA):** tRNA helps decode mRNA into protein by bringing the appropriate amino acids to the ribosome.
- **miRNA (MicroRNA):** miRNAs are small RNA molecules that regulate gene expression by binding to complementary mRNA molecules, leading to mRNA degradation or translation inhibition.

- **lncRNA (Long Non-Coding RNA):** lncRNAs have various regulatory roles, including chromatin remodeling and gene silencing.
- **RNA-Seq:** RNA sequencing has revolutionized the study of RNA, enabling high-throughput and detailed analysis of the entire transcriptome. RNA-Seq data provides information on gene expression levels, the discovery of new RNA species, and insight into alternative splicing and RNA modifications.

Protein Data

Proteins are the molecular machines that perform most cellular functions. They are synthesized based on the information encoded in DNA and processed via RNA. Proteins are composed of amino acids and fold into specific three-dimensional shapes that determine their function. Protein data is critical for understanding how genes influence cellular processes, disease mechanisms, and metabolic pathways.

Key aspects of protein data include:

- **Protein Sequences:** The sequence of amino acids in a protein is directly determined by the mRNA sequence. Protein sequence data is generated by various methods, including mass spectrometry and computational predictions. This data is typically stored in protein databases like **UniProt** and **Protein Data Bank (PDB).**
- **Protein Structure:** The three-dimensional structure of proteins is crucial to understanding their function. Proteins fold into specific shapes that allow them to interact with other molecules, catalyze biochemical reactions, or provide structural support. Structural data is often derived from techniques like **X-ray crystallography**, **NMR spectroscopy**, or **cryo-electron microscopy (cryo-EM).** Understanding protein structure is essential for drug design, enzyme engineering, and studying protein-protein interactions.
- **Post-Translational Modifications (PTMs):** Once proteins are synthesized, they often undergo chemical modifications that can alter their activity, stability, localization, or interactions. PTMs include phosphorylation, acetylation, glycosylation, and ubiquitination, which regulate cellular processes such as signal transduction, DNA repair, and protein degradation.
- **Protein Expression and Abundance:** Data on protein expression levels is used to understand how cells respond to different conditions or stimuli. **Mass spectrometry-based proteomics** can quantify proteins across different conditions, providing insights into disease states, drug responses, and cellular mechanisms. Proteomics also includes the study of protein isoforms and protein modifications.

- **Protein-Protein Interactions (PPIs):** Proteins rarely function alone. Instead, they interact with one another to form complex networks that carry out cellular functions. Studying protein-protein interactions helps in understanding cellular pathways, signaling networks, and disease mechanisms. Techniques like **co-immunoprecipitation (Co-IP), yeast two-hybrid screens**, and **protein microarrays** are commonly used to explore PPIs.

Integration of DNA, RNA, and Protein Data

While each type of biological data provides crucial insights into different aspects of cellular and molecular function, the true power of these data types is realized when they are integrated. DNA data tells us about the genetic potential of an organism, RNA data reveals which genes are being actively expressed at a given time, and protein data informs us about the functional molecules actually performing the work in the cell. By combining these data types, researchers can develop a more comprehensive understanding of biological processes and disease mechanisms, leading to advancements in personalized medicine, biotechnology, and drug discovery. In modern bioinformatics, integrating DNA, RNA, and protein data is central to **systems biology**, which aims to understand how genetic, transcriptomic, and proteomic networks function as a whole. Advanced computational techniques and machine learning algorithms are increasingly used to combine these data sources, making it possible to predict biological outcomes, identify biomarkers, and discover new therapeutic targets.

GENOMIC, TRANSCRIPTOMIC AND PROTEOMIC DATA

Genomic, transcriptomic, and proteomic data are foundational pillars of modern molecular biology, each providing crucial insights into the intricate molecular machinery of life. Genomic data offers the complete blueprint of an organism's genetic makeup, revealing the sequence of DNA that codes for all the functional elements of the genome. Transcriptomic data, on the other hand, provides a snapshot of gene expression, illustrating how genes are activated or silenced in response to various cellular conditions and environmental cues. Proteomic data delves deeper into the functional output of the genome, capturing the complex world of proteins—how they are expressed, modified, interact with each other, and execute the majority of cellular processes. In bioinformatics, these three types of data are often analyzed in concert to gain a more comprehensive understanding of biological systems, diseases, and treatments. By leveraging computational tools and integrative approaches, researchers can uncover the dynamic relationships between genes, RNA, and proteins, helping to decipher the molecular basis of health and disease with unprecedented accuracy and depth. Bioinformatics is a multidisciplinary field that applies computational and statistical techniques to analyze biological data, particularly genetic, transcriptomic, and proteomic data. These types of data are central to understanding the molecular mechanisms of life and disease. Below is an overview of genomic, transcriptomic, and proteomic data and their importance in bioinformatics.

Genomic Data

Genomics refers to the study of the entire genome of an organism, which encompasses the complete set of DNA, including all of its genes. Genomic data typically refers to DNA sequences, including the coding (exons), non-coding regions (introns, promoters, enhancers), and intergenic areas.

Key Features:

- **DNA Sequences:** Genomic data often comes from sequencing technologies such as next-generation sequencing (NGS) and older methods like Sanger sequencing.
- **Variant Detection:** Identifying variations (mutations, SNPs, indels) within a genome that can affect phenotypes.
- **Structural Variations:** Large-scale changes in the genome, such as deletions, duplications, inversions, or translocations.
- **Functional Genomics:** Understanding the roles of genes and regulatory elements in the genome.

Bioinformatics Techniques:

- **Genome Assembly:** Assembling short DNA reads into a complete genome sequence using computational methods.
- **Genome Annotation:** Identifying genes and their functions in a genome, including non-coding RNAs, regulatory regions, and repetitive elements.
- **Comparative Genomics:** Comparing genomes from different species to identify conserved regions, evolutionary relationships, and functional elements.

Transcriptomic Data

Transcriptomics focuses on the study of RNA molecules transcribed from DNA, representing the active gene expression of a cell at a given time. This type of data is crucial for understanding how genes are regulated and how they function in different biological processes.

Key Features:

- **RNA Sequencing (RNA-Seq):** The most common method to obtain transcriptomic data, providing information on the presence, quantity, and structure of RNA molecules.
- **Gene Expression:** Transcriptomics allows for the measurement of gene expression levels across different conditions, tissues, or disease states.
- **Alternative Splicing:** Identification of different mRNA isoforms generated by alternative splicing of a single gene.
- **Non-coding RNAs:** Analysis of non-coding RNA species (miRNAs, lncRNAs, etc.) that play regulatory roles.

Bioinformatics Techniques:

- **Differential Expression Analysis:** Identifying genes that are upregulated or downregulated between different conditions.
- **Gene Ontology (GO) Enrichment:** Understanding the biological functions associated with differentially expressed genes.
- **Transcriptome Assembly:** Constructing transcripts from RNA-Seq data, even in the absence of a reference genome.
- **Pathway Analysis:** Linking differentially expressed genes to biological pathways (e.g., KEGG pathways).

Proteomic Data

Proteomics is the study of the entire set of proteins expressed by an organism, system, or cell at a given time. Proteomic data provides insights into the functional components of cells, as proteins are the key mediators of most cellular functions.

Key Features:

- **Protein Identification:** Proteomic data often comes from mass spectrometry (MS), which allows identification and quantification of proteins present in a sample.
- **Post-translational Modifications (PTMs):** Proteins undergo various modifications after translation (e.g., phosphorylation, acetylation, ubiquitination) that can alter their function.
- **Protein-Protein Interactions (PPI):** Studying how proteins interact with one another to form complexes and contribute to cellular processes.
- **Quantitative Proteomics:** Measuring the abundance of proteins under different conditions or in response to a treatment.

Bioinformatics Techniques:

- **Protein Database Search:** Matching mass spectrometry data to known protein sequences in databases like UniProt or NCBI.
- **Quantitative Analysis:** Using techniques such as label-free quantification or isotope labeling (e.g., SILAC, TMT) to compare protein abundances across conditions.
- **Protein Function Prediction:** Using sequence homology and structural predictions to infer the functions of uncharacterized proteins.
- **Interactome Analysis:** Identifying and analyzing the network of protein-protein interactions.

Integrating Genomic, Transcriptomic, and Proteomic Data

Integrating these three types of data (often referred to as multi-omics) provides a more comprehensive understanding of biological systems. By combining genomic, transcriptomic, and proteomic information, bioinformaticians can:

- Investigate how genetic variations affect gene expression and, in turn, protein expression and function.
- Identify biomarkers for diseases or drug responses at the genetic, transcript, and protein levels.
- Study complex diseases by looking at the interplay between genes, RNA molecules, and proteins.

Techniques for Integration:

- **Systems Biology Approaches:** These approaches use computational models to understand the dynamic interactions between genes, RNA, and proteins.
- **Correlation Analysis:** Comparing changes in gene expression (transcriptomics) with corresponding protein abundance (proteomics) to identify regulatory relationships.
- **Network Analysis:** Mapping genes, proteins, and their interactions onto biological networks to understand cellular functions and disease mechanisms.

Applications in Bioinformatics:

- **Personalized Medicine:** Genomic, transcriptomic, and proteomic data can be used to tailor medical treatments to individual patients based on their molecular profiles.

- **Cancer Research:** These data types are crucial for identifying cancer-specific mutations, gene expression signatures, and protein changes associated with different cancer types.
- **Drug Discovery:** Integrating multi-omics data can help identify new drug targets, predict drug responses, and monitor treatment effectiveness.
- **Functional Genomics:** Understanding how genetic variants influence disease processes and biological functions.

In bioinformatics, genomic, transcriptomic, and proteomic data are essential for gaining a comprehensive understanding of biological systems. Each type of data offers a different perspective—genomic data provides the blueprint, transcriptomic data reveals the active gene expression, and proteomic data shows how these instructions are carried out in the form of proteins. The integration of these data types enables a deeper, more holistic understanding of cellular processes and disease mechanisms, making it an invaluable resource in research and clinical applications.

BIOLOGICAL DATABASES

Biological databases are central to the field of bioinformatics, serving as invaluable repositories of biological information that support research, analysis, and discovery in genomics, transcriptomics, proteomics, and various other branches of biology. These databases store vast amounts of data, including DNA sequences, protein structures, gene expression profiles, metabolic pathways, and much more. By providing organized access to these data, biological databases allow researchers to efficiently query, analyze, and interpret complex biological information, facilitating the understanding of genetic variations, disease mechanisms, evolutionary relationships, and molecular functions.

The rapid advancement of high-throughput technologies, such as next-generation sequencing (NGS), mass spectrometry, and microarrays, has exponentially increased the amount of biological data generated. In response, biological databases have grown in both size and complexity, evolving from simple sequence storage systems to sophisticated platforms that integrate multi-omics data, including genomic, transcriptomic, proteomic, and metabolomic information. These databases enable researchers to not only explore individual biological entities like genes or proteins but also to investigate their interactions, regulatory networks, and roles in cellular processes.

Types of Biological Databases

- **Genomic Databases:** Genomic databases focus on the storage and analysis of DNA sequences, genomic variations, and genome-wide annotations. One of the most well-known examples is the **GenBank** database, which is maintained by the National Center for Biotechnology Information (NCBI). GenBank contains millions of sequences from a wide range of organisms and provides a powerful tool for sequence retrieval and annotation. Another important genomic database is **Ensembl**, which provides detailed genome annotations, gene models, and comparative genomics resources for a variety of species. The **UCSC Genome Browser** also offers a comprehensive view of annotated genomic sequences, allowing users to visualize genes, regulatory regions, and genetic variations across different species. These databases play a critical role in studies of genomics and functional genetics, providing essential resources for genome assembly, variant calling, and gene annotation.

- **Transcriptomic Databases:** Transcriptomic databases are specialized resources for storing RNA sequences and gene expression data. **The Gene Expression Omnibus (GEO)**, hosted by NCBI, is one of the most widely used public repositories for high-throughput gene expression data, including microarray and RNA-Seq datasets. GEO allows researchers to search for gene expression profiles associated with specific conditions, tissues, or diseases, making it a valuable resource for understanding gene function and regulation. **ArrayExpress**, hosted by the European Bioinformatics Institute (EBI), is another major resource for transcriptomic data, particularly for microarray and RNA-Seq experiments. Both GEO and ArrayExpress serve as central hubs for transcriptomic data, enabling users to perform differential gene expression analysis, identify biomarkers, and explore gene networks.
- **Proteomic Databases:** Proteomic databases store information about proteins, including their sequences, structures, functions, post-translational modifications, and interactions. One of the most widely used proteomic databases is **UniProt**, a comprehensive protein sequence and functional annotation resource. UniProt is divided into two main components: **UniProtKB/Swiss-Prot**, which contains manually curated protein sequences and annotations, and **UniProtKB/TrEMBL**, which includes automatically annotated sequences from genome projects. **Protein Data Bank (PDB)** is another key resource, providing 3D structures of proteins, nucleic acids, and complex biomolecular assemblies. Researchers use the PDB to study the structure-function relationships of proteins, explore protein-ligand interactions, and understand the molecular basis of diseases. In addition to these, **PRIDE** (PRoteomics IDEntifications) is a database focused on storing raw and processed mass spectrometry-based proteomics data, facilitating the analysis of protein identification and quantification across different experimental conditions.

- **Pathway and Interaction Databases**: Understanding how proteins and genes work together within biological pathways and networks is crucial for unraveling cellular functions and disease mechanisms. Databases like **KEGG (Kyoto Encyclopedia of Genes and Genomes)** provide information on metabolic pathways, signal transduction pathways, and the interactions between genes, proteins, and small molecules. KEGG integrates functional annotations with pathway diagrams, offering an interactive interface for visualizing the interconnectedness of biological systems. **Reactome** is another comprehensive pathway database that focuses on human biological pathways, providing curated pathway maps and analysis tools to study signaling, metabolism, and disease-related pathways. Similarly, **BioGRID** and **STRING** are databases that focus on protein-protein interactions, offering valuable insights into the physical and functional interactions between proteins in cells.
- **Variant Databases:** Genetic variation databases are essential for identifying and understanding the effects of genetic mutations in health and disease. **dbSNP** (Database of Single Nucleotide Polymorphisms), maintained by NCBI, is one of the largest resources for identifying SNPs and small genetic variations in human populations. This database includes both common and rare genetic variants, which can be used in association studies to identify genetic risk factors for diseases. **ClinVar**, also hosted by NCBI, provides information on the clinical significance of genetic variants, including those associated with inherited diseases, cancer, and drug responses. Other specialized databases, such as **COSMIC** (Catalogue of Somatic Mutations in Cancer), focus on cancer-related mutations, while **HGMD** (Human Gene Mutation Database) is dedicated to pathogenic mutations associated with inherited disorders.
- **Integrated Multi-Omics Databases:** As biological research becomes increasingly interdisciplinary, integrated multi-omics databases are emerging to bridge genomic, transcriptomic, proteomic, and other types of data. **The Cancer Genome Atlas (TCGA)** is one such example, offering multi-omics data (genomic, transcriptomic, and proteomic) from a large number of cancer samples. TCGA enables researchers to explore the molecular basis of cancer, identify potential therapeutic targets, and investigate biomarkers for diagnosis and prognosis. Another example is **Pan-Cancer Analysis of Whole Genomes (PCAWG)**, which integrates data from multiple cancer types to uncover common genomic alterations and molecular pathways driving cancer progression.

Importance of Biological Databases in Research

Biological databases are indispensable tools in the modern life sciences, enabling the storage, retrieval, and analysis of vast amounts of data generated by high-throughput experiments. They serve as a critical infrastructure for advancing our understanding of biological processes, disease mechanisms, and therapeutic interventions. Researchers rely on these databases to identify novel genes, analyze gene expression patterns, discover biomarkers for disease, and explore protein functions and interactions. Moreover, biological databases facilitate collaboration by providing a centralized, accessible platform for sharing data across the scientific community. As biological data continues to grow at an exponential rate, the role of databases in enabling integrative, large-scale analyses will become even more crucial. Future advancements in data management, machine learning, and artificial intelligence will likely lead to even more sophisticated databases capable of providing deeper insights into the complexities of biological systems. Ultimately, biological databases are not only repositories of data but also essential tools for driving innovation in genomics, drug discovery, and personalized medicine.

3 Chapter

BASIC CONCEPTS IN MOLECULAR BIOLOGY

INTRODUCTION

Molecular biology is a branch of science that focuses on understanding the molecular mechanisms behind biological processes. It primarily deals with the structure and function of macromolecules—such as DNA, RNA, and proteins—that are essential for life. At the heart of molecular biology is the concept of the central dogma, which describes the flow of genetic information within a biological system: DNA is transcribed into RNA, and RNA is translated into proteins. This fundamental process underpins cellular functions and is crucial for everything from cellular growth and division to the regulation of metabolic pathways. DNA, composed of nucleotides, carries genetic information in the form of genes, which are sequences that code for proteins. Proteins, in turn, perform a vast array of functions, from catalyzing biochemical reactions (enzymes) to serving as structural components of cells and tissues. The central dogma is not a one-way process; feedback loops and post-transcriptional and post-translational modifications add layers of complexity to gene expression regulation. Techniques like polymerase chain reaction (PCR), gel electrophoresis, and DNA sequencing have revolutionized molecular biology, allowing for the manipulation and analysis of genetic material in ways that were once impossible. Furthermore, molecular biology has profound implications for fields like medicine, biotechnology, and genetics, enabling advancements in areas like gene therapy, genetic engineering, and personalized medicine. The interplay between molecular biology, genomics, and bioinformatics continues to drive discoveries that deepen our understanding of life at the molecular level.

DNA, RNA AND PROTEIN STRUCTURE

Nucleic acids are the organic materials present in all organisms in the form of DNA or RNA. These nucleic acids are formed by the combination of nitrogenous bases, sugar molecules and phosphate groups that are linked by different bonds in a series of sequences. The DNA structure defines the basic genetic makeup of our body. In fact, it defines the genetic makeup of nearly all life on earth. DNA, RNA, and proteins are fundamental macromolecules in biology, each playing a crucial role in storing, transmitting, and executing genetic information within cells. Understanding their structure is key to understanding their functions in the cell.

DNA (DEOXYRIBONUCLEIC ACID) STRUCTURE

DNA is a group of molecules that is responsible for carrying and transmitting the hereditary materials or the genetic instructions from parents to offsprings. This is also true for viruses, as most of these entities have either RNA or DNA as their genetic material. For instance, some viruses may have RNA as their genetic material, while others have DNA as the genetic material. The Human Immunodeficiency Virus (HIV) contains RNA, which is then converted into DNA after attaching itself to the host cell. DNA is known as Deoxyribonucleic Acid. It is an organic compound that has a unique molecular structure. It is found in all prokaryotic cells and eukaryotic cells. DNA was first recognized and identified by the Swiss biologist Johannes Friedrich Miescher in 1869 during his research on white blood cells. The double helix structure of a DNA molecule was later discovered through the experimental data by James Watson and Francis Crick. Finally, it was proved that DNA is responsible for storing genetic information in living organisms.

The following diagram explains the DNA structure representing the different parts of the DNA. DNA comprises a sugar-phosphate backbone and the nucleotide bases (guanine, cytosine, adenine and thymine).

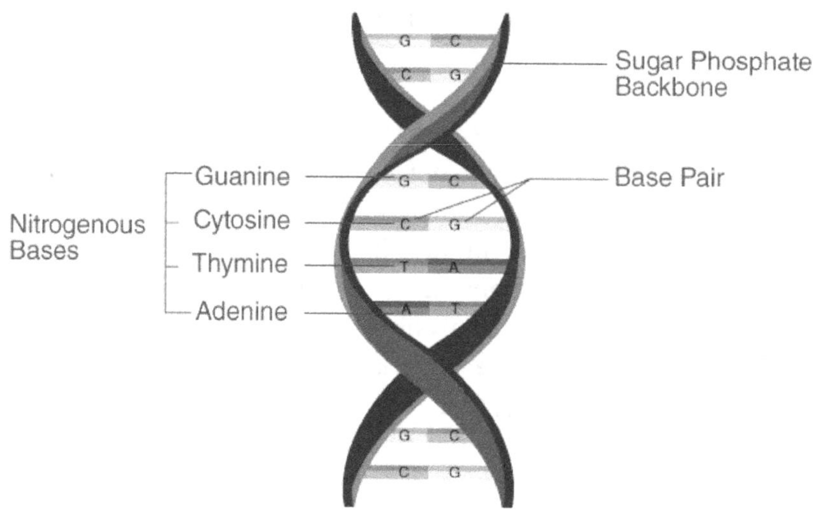

DNA Diagram representing the DNA Structure

Apart from being responsible for the inheritance of genetic information in all living beings, DNA also plays a crucial role in the production of proteins. Nuclear DNA is the DNA contained within the nucleus of every cell in a eukaryotic organism. It codes for the majority of the organism's genomes while the mitochondrial DNA and plastid DNA handles the rest.

The DNA present in the mitochondria of the cell is termed mitochondrial DNA. It is inherited from the mother to the child. In humans, there are approximately 16,000 base pairs of mitochondrial DNA. Similarly, plastids have their own DNA, and they play an essential role in photosynthesis.

DNA is a long, double-stranded helical molecule composed of nucleotides. Each nucleotide consists of three components:

- **A phosphate group** – a negatively charged molecule that links nucleotides together through phosphodiester bonds.
- **A sugar molecule** – deoxyribose, a five-carbon sugar, which forms the backbone of the DNA strand.
- **A nitrogenous base** – there are four types of nitrogenous bases in DNA: adenine (A), thymine (T), cytosine (C), and guanine (G). These bases pair specifically: adenine with thymine, and cytosine with guanine, through hydrogen bonds. This complementary base pairing holds the two strands of the DNA double helix together.

The two strands of DNA run in opposite directions, and their helical shape is stabilized by base pairing and the twisting of the sugar-phosphate backbone. The specific sequence of these nitrogenous bases encodes genetic information.

The DNA structure can be thought of as a twisted ladder. This structure is described as a double-helix, as illustrated in the figure above. It is a nucleic acid, and all nucleic acids are made up of nucleotides. The DNA molecule is composed of units called nucleotides, and each nucleotide is composed of three different components such as sugar, phosphate groups and nitrogen bases.

The basic building blocks of DNA are nucleotides, which are composed of a sugar group, a phosphate group, and a nitrogen base. The sugar and phosphate groups link the nucleotides together to form each strand of DNA. Adenine (A), Thymine (T), Guanine (G) and Cytosine (C) are four types of nitrogen bases.

These 4 Nitrogenous bases pair together in the following way: **A** with **T,** and **C** with **G.** These base pairs are essential for the DNA's double helix structure, which resembles a twisted ladder.

The order of the nitrogenous bases determines the genetic code or the DNA's instructions.

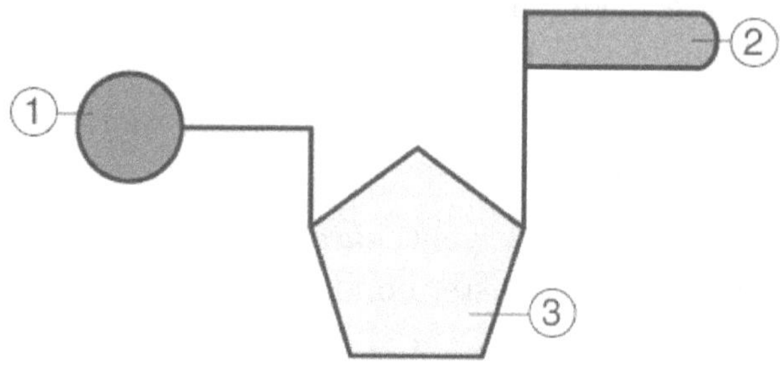

❶ Phosphate | ❷ Nitrogenous Base | ❸ Deoxyribose Sugar

Components of DNA Structure

Among the three components of DNA structure, sugar is the one which forms the backbone of the DNA molecule. It is also called deoxyribose. The nitrogenous bases of the opposite strands form hydrogen bonds, forming a ladder-like structure.

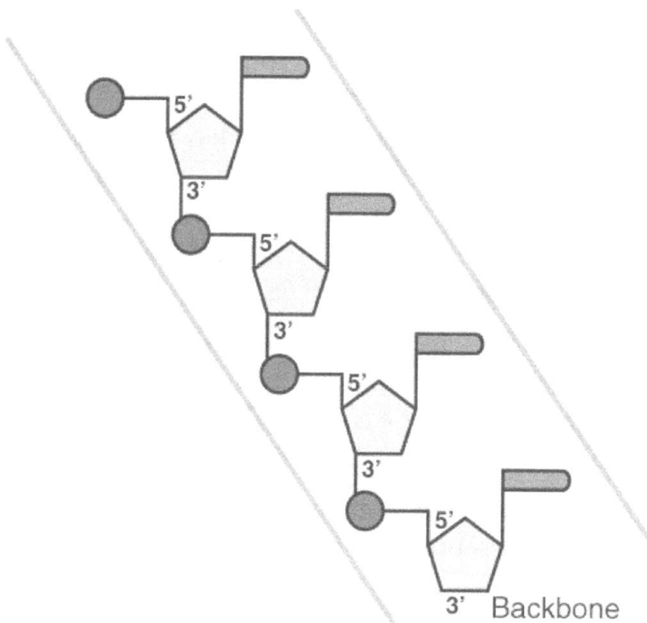

DNA Structure Backbone

The DNA molecule consists of 4 nitrogen bases, namely adenine (A), thymine (T), cytosine (C) and Guanine (G), which ultimately form the structure of a nucleotide. The A and G are purines, and the C and T are pyrimidines.

The two strands of DNA run in opposite directions. These strands are held together by the hydrogen bond that is present between the two complementary bases. The strands are helically twisted, where each strand forms a right-handed coil, and ten nucleotides make up a single turn.

The pitch of each helix is 3.4 nm. Hence, the distance between two consecutive base pairs (i.e., hydrogen-bonded bases of the opposite strands) is 0.34 nm.

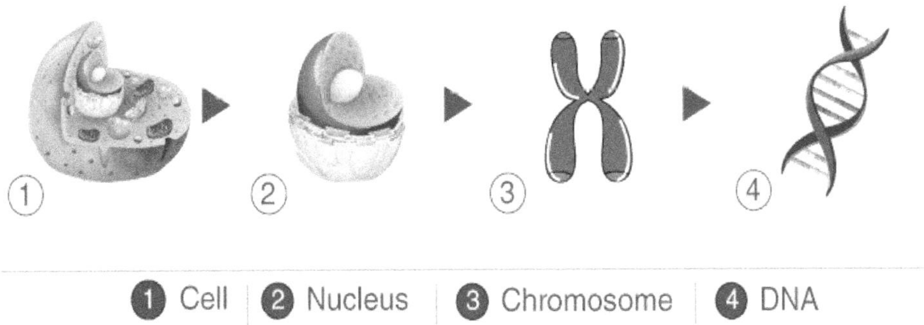

The DNA coils up, forming chromosomes, and each chromosome has a single molecule of DNA in it. Overall, human beings have around twenty-three pairs of chromosomes in the nucleus of cells. DNA also plays an essential role in the process of cell division.

RNA (RIBONUCLEIC ACID) STRUCTURE

RNA is a ribonucleic acid that helps in the synthesis of proteins in our body. This nucleic acid is responsible for the production of new cells in the human body. It is usually obtained from the DNA molecule. RNA resembles the same as that of DNA, the only difference being that it has a single strand unlike the DNA which has two strands and it consists of an only single ribose sugar molecule in it. Hence is the name Ribonucleic acid. RNA is also referred to as an enzyme as it helps in the process of chemical reactions in the body. RNA is similar to DNA but with key differences. The ribonucleic acid has all the components same to that of the DNA with only 2 main differences within it. RNA has the same nitrogen bases called the adenine, Guanine, Cytosine as that of the DNA except for the Thymine which is replaced by the uracil. Adenine and uracil are considered as the major building blocks of RNA and both of them form base-pair with the help of 2 hydrogen bonds. RNA resembles a hairpin structure and like the nucleotides in DNA, nucleotides are formed in this ribonucleic material(RNA). Nucleosides are nothing but the phosphate groups which sometimes also helps in the production of nucleotides in the DNA.

RNA is usually single-stranded and consists of nucleotides made up of:

- **A phosphate group**
- **A sugar molecule** – ribose, which differs from deoxyribose by having a hydroxyl group (-OH) at the 2' carbon, instead of just a hydrogen atom.

- **A nitrogenous base** – RNA contains four bases: adenine (A), uracil (U), cytosine (C), and guanine (G). Unlike DNA, RNA uses uracil in place of thymine, and it pairs adenine with uracil, and cytosine with guanine.

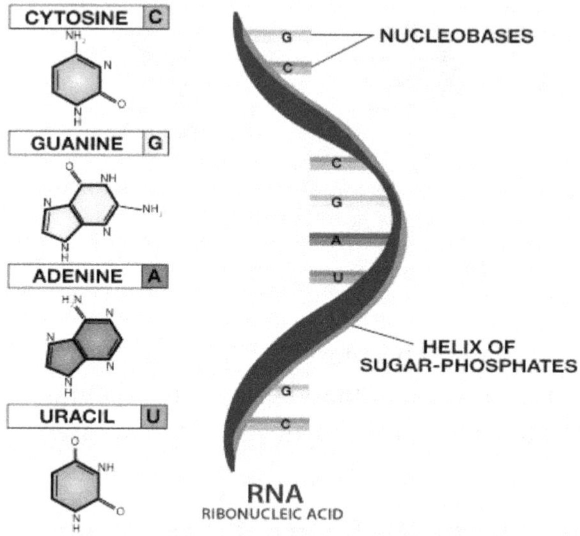

RNA serves as a temporary copy of the genetic information stored in DNA and plays a central role in the synthesis of proteins (via messenger RNA, mRNA) and regulation of gene expression (via other types of RNA like ribosomal RNA, rRNA, and transfer RNA, tRNA).

Functions of RNA

The ribonucleic acid – RNA, which are mainly composed of nucleic acids, are involved in a variety of functions within the cell and are found in all living organisms including bacteria, viruses, plants, and animals. These nucleic acid functions as a structural molecule in cell organelles and are also involved in the catalysis of biochemical reactions. The different types of RNA are involved in various cellular process. The primary functions of RNA:

- Facilitate the translation of DNA into proteins
- Functions as an adapter molecule in protein synthesis
- Serves as a messenger between the DNA and the ribosomes.
- They are the carrier of genetic information in all living cells
- Promotes the ribosomes to choose the right amino acid which is required in building up of new proteins in the body.

RNA Types

There are various types of RNA, out which most well-known and most commonly studied in the human body are:

- **tRNA – Transfer RNA:** The transfer RNA is held responsible for choosing the correct protein or the amino acids required by the body in-turn helping the ribosomes. It is located at the endpoints of each amino acid. This is also called as soluble RNA and it forms a link between the messenger RNA and the amino acid.
- **rRNA-Ribosomal RNA:** The rRNA is the component of the ribosome and are located within the in the cytoplasm of a cell, where ribosomes are found. In all living cells, the ribosomal RNA plays a fundamental role in the synthesis and translation of mRNA into proteins. The rRNA is mainly composed of cellular RNA and are the most predominant RNA within the cells of all living beings.
- **mRNA – Messenger RNA.:** This type of RNA functions by transferring the genetic material into the ribosomes and pass the instructions about the type of proteins, required by the body cells. Based on the functions, these types of RNA is called the messenger RNA. Therefore, the mRNA plays a vital role in the process of transcription or during the protein synthesis process.

Protein Structure

Proteins are complex molecules made up of amino acids, which are the building blocks of proteins. The structure of proteins can be described at four levels:

- **Primary Structure** – The linear sequence of amino acids in a polypeptide chain. The sequence is determined by the gene encoding the protein and is crucial to its final shape and function.
- **Secondary Structure** – The local folding of the polypeptide chain into regular structures such as alpha helices and beta sheets. These structures are stabilized by hydrogen bonds between the peptide backbone.
- **Tertiary Structure** – The overall three-dimensional shape of the protein, which is formed by the folding of the secondary structures. Tertiary structure is stabilized by various interactions, including hydrogen bonds, ionic bonds, van der Waals forces, and disulfide bridges (covalent bonds between sulfur atoms in cysteine residues).
- **Quaternary Structure** – The arrangement of multiple polypeptide chains (subunits) in a multi-subunit protein. Not all proteins have quaternary structure, but for those that do, the subunits work together to form the functional protein.

The unique sequence of amino acids in a protein determines its final structure and, consequently, its function. Proteins can act as enzymes (catalyzing biochemical reactions), structural components (e.g., collagen in connective tissues), transporters (e.g., hemoglobin), and many other roles in the cell.

Relationship Between DNA, RNA, and Protein:

The central dogma of molecular biology describes the flow of genetic information from DNA to RNA to protein:

- DNA holds the genetic blueprint of an organism.
- RNA is synthesized from DNA (transcription) and carries the instructions to the ribosome for protein synthesis.
- Proteins are synthesized from RNA templates (translation) in the ribosome, where the mRNA is read in sets of three bases (codons) that correspond to specific amino acids.

Thus, the structure of DNA determines the genetic code, which is transcribed into RNA, and ultimately translated into the functional proteins that carry out the biological functions of the cell.

CENTRAL DOGMA AND MOLECULAR BIOLOGY

A Central Dogma is an explanation of the flow of Genetic information in a cell, including the replication of the DNA, the transcription of the RNA, and the translation of the RNA to create the Proteins. It is possible to understand the concept of interaction through the framework. For example, biopolymers are common. Biopolymers are primarily composed of Proteins, RNA, and DNA, which can further be divided into general and unknown transfers. In the laboratory, special transfers are made in exceptional cases. Cells are generally transferred by general means. Information is constantly flowing through transcription and translation in the human body. It is believed that unknown transfers never occur. As a result, two new strands of DNA are created, one strand from the parent DNA and one from the newly synthesized DNA. This is called semiconservative DNA replication.

The sum total of all the DNA in an organism is called its genome. Genomic information is like a computer program for a cell. If DNA is the master copy of a cell's genetic program, its integrity must be preserved. One way the DNA is protected is because RNA acts as the working copy. Chemically, RNA is very similar to DNA. Biochemically, the major difference is that RNA either acts as a component of the metabolic machinery or is a copy of the information for protein synthesis. The relationship between DNA and RNA is called the central dogma of molecular biology. The central dogma of molecular biology describes the flow of information from DNA through RNA into proteins. This flow of information is called gene expression. It occurs through two main processes: transcription and translation.

DNA → RNA → protein

DNA contains genes that code for proteins. RNA is the intermediate between DNA and proteins. It carries information in genes from the nucleus to the cytoplasm in eukaryotes. Proteins determine the structure and the function of a particular cell. A protein is composed of an amino acid sequence, which is the coding sequence of a gene. Gene expression is the process of synthesizing proteins based on the instructions in genes. The two steps of gene expression are transcription and translation. The ribosome translates the nucleotide sequence of the mRNA into an amino acid polymer using tRNAs to translate triplets of nucleotides into the amino acids specified by the genetic code.

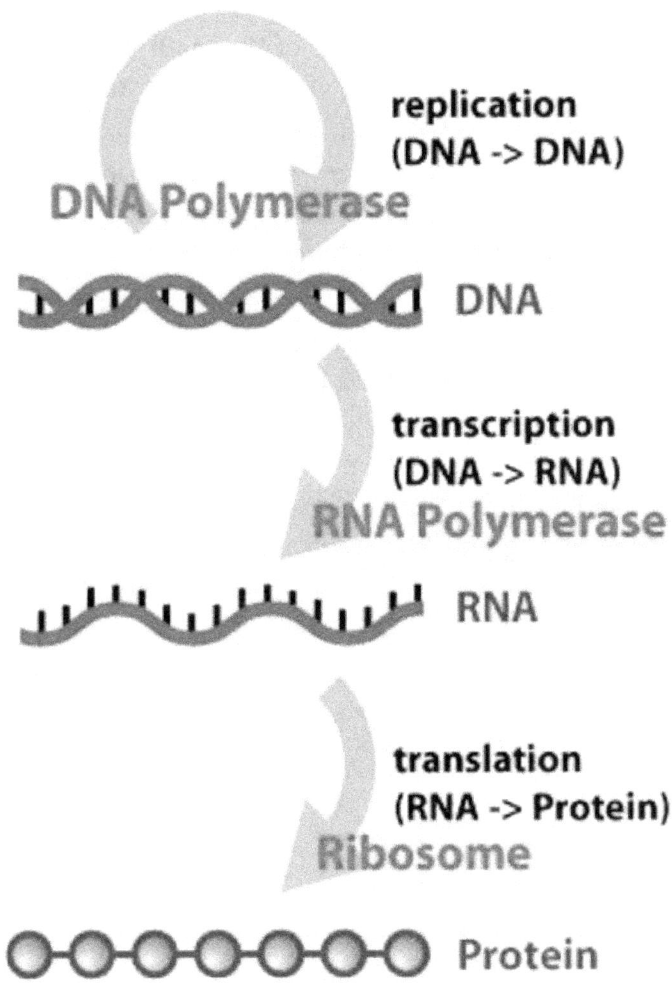

Central Dogma of Molecular Biology

Transcription is the synthesis of an RNA molecule that contains the coding sequence of a gene. The enzyme RNA polymerase recognizes template promoter sequences and initiates, elongates and terminates the synthesis of an unstable intermediate RNA molecule In some cases this RNA is functional (i.e. rRNAs, tRNAs) but in most cases it is a transient message due to RNA's inherent instability. The primary transcript in prokarytoes is the mature mRNA. In eukaryotes several post-transcriptional modifications - including addition of the 5' CAP, the removal of introns and joining of exons to form a contiguous coding sequence, and cleavage and addition of the 3' poly-A tail are observed. Translation follows transcription in which the amino acid sequence of a gene is synthesized based on the coding sequence in mRNA. As translation proceeds, the newly synthesized protein folds through a series of transient states to eventually achieve its mature 3D shape. The protein represents the functional expression of the infomation stored in the heritable storehouse. The surfaces of the mature protein are responsible for molecular recognition and signaling, enzymatic and structural functions.

In addition to the universal flow of information from DNA to RNA to proteins, some alternative mechanisms occur in different types of organisms. Reverse transcription, RNA replication, and Direct translation of DNA into proteins are three such unusual flows of information.

- **Reverse Transcription:** Transferring the information of RNA into DNA occurs during the reverse transcription process. It mainly occurs in retroviruses such as HIV. Also, reverse transcription occurs in retrotransposons and during telomere synthesis in eukaryotes. After reverse transcription, the information flow as usual from cDNA to RNA to Proteins.
- **RNA Replication:** Copying of the RNA information into another RNA happens during the RNA replication process. The enzyme involved in RNA replication is RNA-dependent RNA polymerase. It occurs during RNA silencing and RNA editing in eukaryotes.
- **Direct Translation:** Eukaryotic ribosomes can synthesize proteins from single-stranded DNA in vitro. Figure 2 shows the three types of unusual flow of information in green arrows.

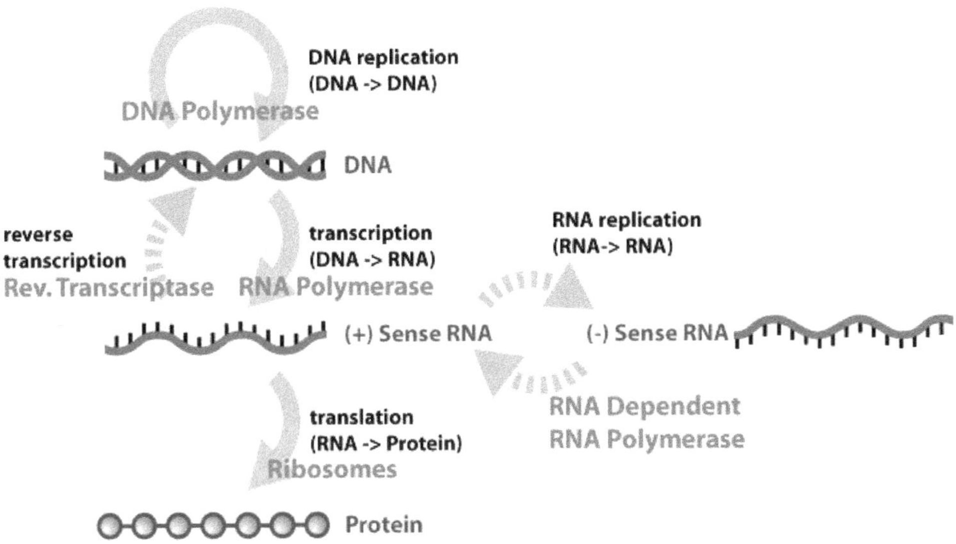

Unusual Flow of Information

The central dogma of molecular biology describes the flow of information from DNA to RNA to proteins. Proteins synthesis is the mechanism of gene expression. It occurs through the transcription of DNA into RNA and translation of RNA into proteins.

GENES, GENOMES AND PROTEINS

Genes, genomes, and proteins are the fundamental pillars of molecular biology, intricately linked in the processes that govern life at the cellular level. Genes, the functional units of heredity, are stretches of DNA that carry the instructions for producing proteins, which in turn drive the vast majority of cellular functions. These instructions are encoded in the sequence of nucleotides (adenine, thymine, cytosine, and guanine) that make up DNA. The complete set of genetic material in an organism—the genome—contains not only the genes themselves but also non-coding regions that regulate gene activity and contribute to the structural organization of the genome. The genome serves as a biological blueprint, ensuring the transmission of genetic information from one generation to the next, while also providing the necessary diversity for evolution. Proteins, the molecular machines of the cell, are synthesized according to the instructions in the genes. They are made up of amino acids and serve a multitude of roles, from catalyzing biochemical reactions as enzymes to forming structural components and facilitating communication within and between cells. The relationship between genes, genomes, and proteins is a dynamic flow of information: genes are transcribed into messenger RNA, which is then translated into proteins, which perform essential functions in the organism. This intricate system of genetic information processing is the key to the development, functioning, and evolution of all living organisms, and understanding it has far-reaching implications in fields like medicine, biotechnology, and evolutionary biology.Genes, genomes, and proteins are the cornerstone of molecular biology, each playing a fundamental role in the storage, expression, and function of genetic information in living organisms.

Genes

A gene is a segment of DNA that contains the instructions for producing a specific protein or functional RNA molecule. Genes are the basic units of heredity and are inherited from parent to offspring. Each gene is made up of a sequence of nucleotides (adenine, thymine, cytosine, and guanine) that codes for a particular trait or function. The sequence of nucleotides within a gene is transcribed into messenger RNA (mRNA), which then undergoes translation to produce proteins. In eukaryotic organisms, genes are often interrupted by non-coding regions known as introns, with the coding sequences (exons) spliced together to form the final mRNA product. The mRNA is then translated by ribosomes in the cytoplasm into a corresponding protein. Genes not only encode proteins but also regulate cellular processes such as growth, division, and response to environmental stimuli. In some cases, genes encode non-coding RNA molecules (such as rRNA and tRNA) that play roles in protein synthesis or gene regulation.

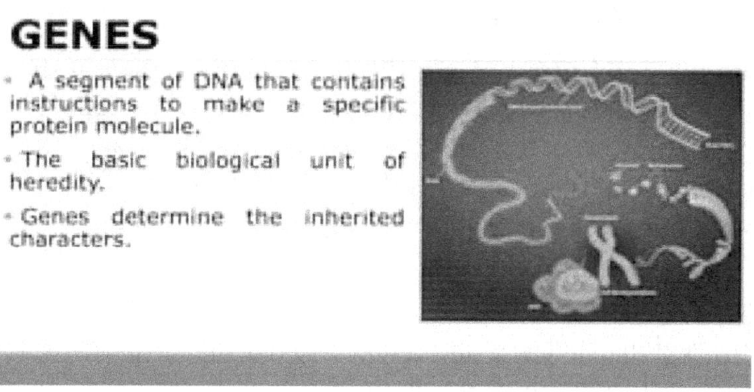

Genomes

The genome is the complete set of genetic material (DNA) in an organism. It includes all of the organism's genes, as well as non-coding sequences of DNA that regulate gene expression and contribute to the structural organization of the genome. Genomes can vary greatly in size and complexity depending on the organism. For example, humans have about 3 billion base pairs of DNA arranged into 23 pairs of chromosomes, while simpler organisms like bacteria may have much smaller genomes. The genome carries all the information necessary for the growth, development, and functioning of an organism. It is the blueprint from which all cellular activities are directed, from the replication of cells to the production of proteins and regulation of metabolism. Genomes are inherited from one generation to the next and undergo changes over time through mutations, which can lead to genetic diversity and evolution. In modern biology, sequencing the genome (genomics) has become a key tool for understanding the molecular basis of diseases, as well as for studying evolution and biodiversity. The human genome project, for instance, successfully mapped the entire human genome, providing immense insights into human biology and health.

Proteins

Proteins are large, complex molecules that are essential for nearly all biological processes. They are made up of amino acids, and their specific sequence of amino acids (encoded by genes) determines their three-dimensional structure and function. Proteins play a vast array of roles in the cell, including:

- **Enzymes:** Proteins that catalyze biochemical reactions, speeding up metabolic processes.
- **Structural proteins**: These provide support and shape to cells and tissues (e.g., collagen in connective tissues).
- **Transport proteins**: These move molecules across membranes or through the bloodstream (e.g., hemoglobin transporting oxygen in the blood).
- **Signaling proteins**: These are involved in communication between cells, tissues, and organs (e.g., hormones like insulin).
- **Regulatory proteins:** These help regulate gene expression and control cellular functions.

The synthesis of proteins from genetic information involves two key processes:

- **Transcription:** The DNA sequence of a gene is transcribed into mRNA, which serves as a messenger between the DNA in the nucleus and the ribosome in the cytoplasm.
- **Translation:** The mRNA sequence is translated into a polypeptide chain of amino acids by the ribosome, which folds into a functional protein.

The structure of proteins is hierarchical, beginning with the primary sequence of amino acids and progressing to higher levels of organization (secondary, tertiary, and quaternary structures). The final shape of the protein is critical to its function; even slight changes in the amino acid sequence (mutations) can lead to functional changes or disease.

Interrelationship Between Genes, Genomes, and Proteins

- **Genes** are segments of DNA that contain the information to make proteins.
- **Genomes** are the entire set of genetic material (DNA) in an organism, including all the genes.
- **Proteins** are the functional products produced through the processes of transcription and translation from the genetic instructions encoded in genes.

Genes within the genome dictate the production of proteins, which in turn govern the structure and function of cells and organisms. This flow of information from genes to proteins is central to understanding how biological systems operate, how traits are inherited, and how genetic mutations can affect health and development. Through techniques like genome sequencing, gene expression profiling, and protein analysis, scientists are continually advancing our understanding of the molecular biology of life.

4 Chapter

INTRODUCTION TO COMPUTATIONAL BIOLOGY

INTRODUCTION

Computational biology is an interdisciplinary field that leverages the power of computational techniques and algorithms to solve complex biological problems. By integrating concepts from computer science, mathematics, and biology, computational biology allows scientists to analyze and interpret vast amounts of biological data. The rapid advancements in high-throughput technologies, such as next-generation sequencing, have generated an overwhelming influx of biological information, including genomic, transcriptomic, proteomic, and metabolomic data. Computational biology provides the tools necessary to process, store, and analyze this data to uncover new insights into the molecular mechanisms of life, disease, and evolution. At its core, computational biology seeks to develop models and algorithms that can simulate biological processes, predict the behavior of biological systems, and make sense of large, multidimensional datasets. This approach is crucial for understanding complex phenomena such as gene regulation, protein folding, cellular signaling, and the interactions within ecological systems. The field spans a wide range of sub-disciplines, including bioinformatics (focused on the storage and analysis of biological data), systems biology (focused on the modeling of biological networks), and structural biology (which studies the 3D structure of biomolecules using computational techniques). By combining computational tools with biological knowledge, researchers are able to identify novel drug targets, unravel the genetic basis of diseases, and predict the outcomes of experimental interventions, thereby driving advancements in personalized medicine, biotechnology, and environmental science. The intersection of biology and computation also holds promise for addressing some of the grand challenges in biology, such as understanding the intricate regulatory networks that control cellular behavior, mapping the human microbiome, and exploring the potential of artificial intelligence to accelerate scientific discovery. As biological datasets continue to grow in scale and complexity, computational biology will play an increasingly critical role in transforming biology into a data-driven, predictive science. Through the synergy of computational power and biological insight, this field is poised to revolutionize our understanding of life at every level, from the molecular to the ecological.Computational biology is a rapidly growing field that applies mathematical models, computational techniques, and algorithms to understand biological systems, processes, and data. By combining principles from computer science, biology, physics, and mathematics, it enables researchers to model, analyze, and interpret the complex and vast amounts of data generated by modern biological research. The field aims to provide deeper insights into the mechanisms of life, solve biological questions, and support applications across medicine, biotechnology, and environmental science.

The Role of Computational Biology in Modern Science

The explosion of biological data in recent decades has been largely driven by advancements in high-throughput technologies such as next-generation sequencing (NGS), microarrays, and mass spectrometry. These technologies enable the collection of enormous datasets covering various biological domains, including genomics, transcriptomics, proteomics, and metabolomics. However, traditional biological analysis techniques cannot handle the volume, complexity, and multidimensional nature of these datasets. Computational biology offers a way to process, analyze, and interpret these data to derive meaningful biological insights. For example, computational tools allow scientists to:

- **Analyze Genomic Data**: Sequencing technologies produce vast amounts of DNA sequence data, which computational biologists use to assemble genomes, identify genes, and explore genetic variation across individuals or species.
- **Analyze Protein Function and Structure**: Proteomics data allows computational approaches to predict protein structures, understand protein-protein interactions, and model how proteins function in cellular processes.
- **Model Biological Processes**: Computational models of metabolic, genetic, and signaling networks help simulate biological processes, enabling predictions about cellular behavior under various conditions.

Key Areas within Computational Biology

Computational biology is a diverse field with several sub-disciplines that focus on different aspects of biological data and processes:

- **Bioinformatics:** Bioinformatics is one of the most prominent subfields of computational biology. It focuses on the storage, retrieval, and analysis of biological data, particularly genomic data. The development of algorithms and databases to manage large biological datasets is a core part of bioinformatics. Tools such as BLAST (Basic Local Alignment Search Tool) and FASTA are commonly used for sequence alignment, while databases like GenBank and Ensembl store genomic sequences and annotations. Bioinformatics also involves the development of pipelines and workflows for analyzing sequencing data, including genome assembly, variant detection, gene expression analysis, and functional annotation.

- **Systems Biology:** Systems biology takes a holistic approach to understand the interactions within biological systems. Rather than focusing on individual molecules or components, systems biology aims to model how different components (e.g., genes, proteins, metabolites) interact and influence each other in a network. Computational models in systems biology help researchers simulate and predict cellular behaviors in response to various stimuli, contributing to a deeper understanding of cellular processes like signal transduction, gene regulation, and metabolic control. These models can also be used to predict the effects of drug treatments on a system, which is crucial for fields such as drug discovery and personalized medicine.
- **Structural Bioinformatics:** Structural bioinformatics focuses on understanding the three-dimensional structure of biomolecules, especially proteins and nucleic acids. The 3D structure of a molecule is crucial to understanding its function, and computational methods like molecular docking, molecular dynamics simulations, and homology modeling allow scientists to predict and analyze molecular structures. This is especially important in drug design, where understanding how small molecules interact with target proteins can guide the development of new therapeutic compounds.
- **Computational Evolutionary Biology:** This subfield applies computational methods to study evolutionary processes. Techniques like phylogenetic tree construction, molecular evolution modeling, and the analysis of evolutionary patterns across species provide insights into how life has evolved over time. Computational evolutionary biology also plays a crucial role in studying genetic diversity, speciation, and adaptation, as well as understanding the molecular basis of diseases like cancer, where mutations accumulate over time.

Techniques and Methods in Computational Biology

Computational biology uses a wide range of mathematical and computational methods to analyze biological data. These techniques can be broadly categorized into:

- **Sequence Analysis:**One of the foundational tasks in computational biology is sequence analysis, which involves comparing and aligning biological sequences (DNA, RNA, or protein) to identify similarities, differences, and evolutionary relationships. Common methods include:
    - **Sequence Alignment**: Aligning sequences to detect conserved regions or mutations. Tools like BLAST and Clustal Omega are commonly used.
    - **Motif Search**: Identifying conserved sequence patterns (motifs) that are likely to have biological significance, such as transcription factor binding sites or RNA secondary structures.
    - **Genome Assembly**: Reconstructing a genome from short DNA sequences generated by sequencing technologies, a task which involves sophisticated algorithms like De Bruijn graphs and overlap-layout-consensus.
- **Statistical Modeling and Machine Learning:** Statistical and machine learning techniques are increasingly applied to the analysis of biological data. For example, machine learning algorithms like support vector machines (SVM), random forests, and deep learning can classify biological data, such as predicting the function of unknown genes or identifying disease-related genetic variants. These methods are also used in drug discovery to predict drug-target interactions or in genomics to identify biomarkers for disease.
- **Molecular Dynamics Simulations:** Molecular dynamics (MD) simulations use principles of physics to model the behavior of atoms and molecules over time. MD simulations are used to predict the structure and dynamics of proteins, nucleic acids, and small molecules in solution. This technique can provide detailed insights into how biomolecules fold, interact, and respond to external factors such as temperature or pH changes, making it an essential tool in structural bioinformatics and drug design.
- **Network Biology and Graph Theory:** Biological systems often function through complex networks of interactions between proteins, genes, and metabolites. Computational biologists use graph theory and network analysis to model these interactions. Biological networks, such as protein-protein interaction (PPI) networks, gene regulatory networks, and metabolic pathways, can be represented as graphs where nodes represent entities (genes, proteins, metabolites) and edges represent interactions. By analyzing these networks, researchers can identify key nodes (e.g., master regulators) and predict how changes in one part of the network might affect the entire system.

Applications of Computational Biology

The power of computational biology lies in its broad applicability to various biological and biomedical fields. Some of the most important applications include:

- **Personalized Medicine:** By analyzing genetic and molecular data from individuals, computational biology plays a central role in personalized medicine, which tailors medical treatment based on an individual's unique genetic makeup. Computational methods can identify genetic variants associated with diseases, predict individual responses to drugs, and help design more effective treatment plans.
- **Drug Discovery and Development:** Computational biology is integral to the drug discovery process, from identifying potential drug targets to screening large databases of compounds for activity against specific biological pathways. In silico drug screening, molecular docking, and molecular dynamics simulations help predict how small molecules will interact with target proteins, accelerating the discovery of new therapeutic agents.
- **Cancer Research:** Computational biology is crucial for understanding the molecular basis of cancer. By analyzing genomic and transcriptomic data from tumor samples, computational tools can identify mutations, gene expression patterns, and signaling pathways that drive cancer progression. This information helps to develop targeted therapies and predictive biomarkers for cancer treatment.
- **Epidemiology and Disease Modeling:** Computational models also play a key role in understanding the spread of infectious diseases. By analyzing epidemiological data and creating models of disease transmission, computational biology aids in predicting outbreaks, evaluating public health strategies, and designing vaccines.

Challenges and Future Directions

While computational biology has revolutionized biological research, several challenges remain. The sheer volume of biological data, particularly from high-throughput technologies, requires powerful computational infrastructure and advanced algorithms for processing and interpretation. Furthermore, integrating data across various levels of biological organization (genomic, transcriptomic, proteomic, etc.) remains a difficult task. There is also a need for more accurate and reliable predictive models to move from correlation to causation, especially in complex diseases like cancer. As artificial intelligence (AI) and machine learning continue to evolve, their integration into computational biology will likely enhance the ability to make sense of massive datasets and uncover new biological knowledge. The advent of quantum computing may also provide breakthroughs in modeling biological processes at a scale and precision that was previously unimaginable. Computational biology represents a frontier in modern biology that combines computational power with biological understanding to solve some of the most pressing problems in science and medicine. As technology continues to evolve, computational biology will remain at the heart of biomedical research, helping to unlock the complexities of life and accelerate innovations in health and biotechnology.

ALGORITHMS IN BIOLOGY

Algorithms and complexity are fundamental concepts in computer science and bioinformatics. Algorithms are step-by-step procedures or formulas for solving a problem, while complexity analysis evaluates the efficiency of algorithms in terms of their time and space requirements. In bioinformatics, algorithms play a crucial role in analyzing biological data, such as DNA sequences, protein structures, and gene expression patterns. These algorithms help researchers identify genes, predict protein structures, and understand evolutionary relationships. Complexity analysis helps researchers evaluate the scalability and efficiency of algorithms. This analysis considers factors such as the size of the input data and the resources required for computation. By understanding algorithm complexity, bioinformaticians can choose the most suitable algorithms for their research, balancing accuracy with computational efficiency.

Algorithms play a critical role in the field of biology, particularly in computational biology and bioinformatics, where vast amounts of biological data are analyzed and interpreted. The application of algorithms in biology allows researchers to gain insights into complex biological systems, identify patterns, make predictions, and even simulate biological processes. These algorithms are based on mathematical and computational principles, and their development has been heavily influenced by advances in both biology and computer science. In this context, algorithms are used to solve a variety of problems, ranging from sequence alignment to structural prediction and from network modeling to machine learning applications.

Types of Algorithms in Biology

There are several categories of algorithms that are particularly relevant in biological research. These include algorithms for data alignment, structural prediction, machine learning, network analysis, and optimization. Below, we will discuss some of the most important algorithms in biology in detail.

Sequence Alignment Algorithms

One of the foundational tasks in bioinformatics is sequence alignment, which involves comparing biological sequences (DNA, RNA, or protein) to identify similarities, differences, or evolutionary relationships. Sequence alignment is central to many biological applications, such as genome assembly, functional annotation, and evolutionary analysis.

- **Pairwise Sequence Alignment:** Pairwise sequence alignment is used to compare two biological sequences to identify their similarities and differences. The two main types of pairwise sequence alignment are **global alignment** and **local alignment**.
    - **Global Alignment**: This type of alignment compares two sequences in their entirety, from start to end. It is typically used when the sequences are similar in length and share substantial regions of homology. The **Needleman-Wunsch algorithm** is a classical dynamic programming approach for global alignment, which aligns the entire length of two sequences while minimizing gaps and mismatches.
    - **Local Alignment**: Local alignment focuses on identifying regions of similarity within two sequences, even if the sequences differ greatly in length. This is useful for detecting conserved motifs or functional regions in otherwise divergent sequences. The **Smith-Waterman algorithm** is used for local sequence alignment and is also based on dynamic programming. It is particularly valuable for comparing parts of sequences that might be conserved across species.
- **Multiple Sequence Alignment (MSA):** Multiple sequence alignment (MSA) extends pairwise alignment to align three or more sequences simultaneously. MSA is critical for studying conserved regions across multiple genes or species. Several algorithms are used for MSA, with the most notable being:
    - **ClustalW**: One of the most widely used algorithms, ClustalW employs a progressive alignment approach, where sequences are progressively aligned based on a guide tree.
    - **MAFFT**: Another popular algorithm for MSA, MAFFT uses various heuristics to achieve faster and more accurate alignments, especially with large datasets.

Sequence Motif Search Algorithms

Motif finding is another crucial task in computational biology. Biological motifs are short, recurring patterns within a biological sequence that often have a biological function, such as transcription factor binding sites or RNA secondary structure elements.

- **Position-Specific Scoring Matrix (PSSM):** A PSSM is a statistical model used to describe sequence motifs. It represents the probability of each nucleotide (or amino acid) occurring at each position in the motif. The **MEME (Multiple Em for Motif Elicitation)** algorithm is one of the most commonly used motif discovery tools, which finds motifs by searching for statistically significant patterns within a set of input sequences.
- **Hidden Markov Models (HMMs):** HMMs are used for motif discovery and for modeling biological sequences where the data is sequential and the output is probabilistic. In the context of sequence motifs, HMMs can be trained on known motifs and then used to identify similar motifs in new sequences. HMMs are used in gene prediction and RNA secondary structure prediction, as well as in protein family classification.

Gene Prediction Algorithms

Gene prediction is a critical task in genomics. The goal is to identify the locations of genes within a genome and predict their structures, including exons, introns, and regulatory regions. This task has become more challenging with the increasing complexity of genomes and the diversity of gene structures across species.

- **GeneMark and Augustus**
  - **GeneMark:** This algorithm is a widely used tool for gene prediction in prokaryotic and eukaryotic genomes. It uses a probabilistic model based on HMMs to predict the most likely gene locations in a given genomic sequence.
  - **Augustus:** Augustus is another popular gene prediction tool that uses both ab initio (from scratch) and evidence-based methods (e.g., transcript data) to predict genes in eukaryotic genomes.

Phylogenetic Tree Construction Algorithms

Phylogenetic tree construction involves the comparison of multiple species or sequences to determine their evolutionary relationships. This task is essential in fields like evolutionary biology, genomics, and comparative genomics.

- **Neighbor-Joining (NJ):** The **Neighbor-Joining** algorithm is a distance-based method for constructing phylogenetic trees. It works by calculating pairwise distances between sequences and then progressively merging the closest sequences into a tree. The method is efficient and works well when the data is relatively straightforward and the evolutionary history is not too complex.

- **Maximum Likelihood (ML) and Bayesian Inference**
  - **Maximum Likelihood** methods involve constructing a phylogenetic tree by maximizing the likelihood that a given tree best represents the observed data. Algorithms like **RAxML** and **PhyML** are widely used for this purpose.
  - **Bayesian Inference** methods use a probabilistic approach to estimate the posterior probability of trees given the data. **MrBayes** is a popular tool for Bayesian phylogenetic analysis.

Structural Bioinformatics Algorithms

Structural bioinformatics is concerned with predicting and analyzing the 3D structures of biological macromolecules, such as proteins and nucleic acids. Predicting how a molecule folds into its three-dimensional shape is crucial for understanding its function and for drug design.

- **Homology Modeling:**Homology modeling predicts the 3D structure of a protein based on its similarity to a known structure. The process involves aligning the target sequence to a homologous template and then using this alignment to build a 3D model. Common algorithms for homology modeling include **MODELLER** and **SWISS-MODEL.**
- **Molecular Docking:** Molecular docking algorithms are used to predict how small molecules (e.g., drugs) bind to target proteins. The goal is to identify the binding affinity and the orientation of the ligand within the receptor's binding site. **AutoDock** and **DockingServer** are two popular docking algorithms that simulate protein-ligand interactions to assist in drug discovery.
- **Molecular Dynamics (MD) Simulations:** Molecular dynamics simulations model the physical movements of atoms and molecules over time. This approach is used to study the stability of protein-ligand complexes, protein folding, and conformational changes. MD simulations use algorithms based on classical mechanics, such as those implemented in **GROMACS** or **AMBER.**

Machine Learning Algorithms in Biology

Machine learning has become an essential tool in many areas of computational biology, from gene expression analysis to disease prediction. By training algorithms on large datasets, machine learning techniques can discover patterns, make predictions, and even uncover novel biological relationships.

- **Support Vector Machines (SVM):** SVM is a supervised learning algorithm used for classification and regression. In bioinformatics, SVMs are often used for tasks like gene classification (e.g., distinguishing between cancerous and normal genes) or predicting protein function based on sequence data.

- **Random Forests:** Random forests are ensemble learning methods that combine the output of multiple decision trees to improve accuracy and robustness. In biology, random forests are used for classifying gene expression data, predicting protein-protein interactions, or identifying disease-associated variants in genomic data.
- **Deep Learning:** Deep learning, particularly convolutional neural networks (CNNs) and recurrent neural networks (RNNs), has seen increasing applications in biology. These algorithms excel at handling unstructured data, such as images and sequences. In biology, deep learning is used for tasks such as protein structure prediction (e.g., **AlphaFold**), sequence motif discovery, and the analysis of genomic data for disease prediction.

Optimization Algorithms in Systems Biology

Optimization algorithms are essential in modeling biological systems, where the goal is to find the best parameters or solutions to a given problem, such as the optimal conditions for metabolic flux or enzyme concentration.

- **Linear Programming:** Linear programming (LP) is widely used in systems biology for optimizing metabolic networks. The **flux balance analysis (FBA)** method, which is based on LP, allows researchers to predict the flow of metabolites through a metabolic network under different conditions.
- **Genetic Algorithms:** Genetic algorithms (GAs) are optimization methods inspired by natural selection. In biological systems, GAs are used to optimize parameters for metabolic models, evolutionary simulations, or protein folding predictions.

The use of algorithms in biology has transformed our ability to understand and manipulate biological systems. From sequencing and alignment to structural prediction and systems biology, algorithms are essential tools that make sense of vast and complex biological data. As computational techniques continue to evolve, the integration of artificial intelligence, machine learning, and optimization algorithms will further advance our ability to solve biological problems, making the intersection of biology and computation an increasingly dynamic and transformative field.

ROLE OF PROGRAMMING IN BIOINFORMATICS

Programming plays a pivotal role in bioinformatics, serving as the backbone for analyzing and interpreting complex biological data. As the field combines biology, computer science, and mathematics, programming enables the development of algorithms, data structures, and computational tools necessary for processing large-scale biological datasets, such as genomic sequences, protein structures, and gene expression profiles. With advancements in high-throughput technologies like next-generation sequencing, bioinformatics has become increasingly reliant on programming skills to efficiently manage and extract meaningful insights from vast amounts of biological information. From sequence alignment to molecular modeling and systems biology, programming is integral to advancing research and discoveries in the life sciences.

Bioinformatics is a multidisciplinary field that combines biology, computer science, mathematics, and statistics to analyze and interpret large volumes of biological data. With the rapid advancements in high-throughput technologies such as next-generation sequencing (NGS), mass spectrometry, and microarrays, the generation of biological data has grown exponentially, necessitating the use of computational methods to manage, analyze, and derive meaningful conclusions from this data. Programming plays a crucial role in bioinformatics by providing the tools and techniques required to handle, manipulate, and analyze this vast and complex biological data.

Data Acquisition and Processing

One of the primary roles of programming in bioinformatics is in the acquisition, cleaning, and preprocessing of biological data. Raw biological data from experimental technologies, such as genomic sequencing or protein identification, are often messy, noisy, and inconsistent. Programming languages like Python, R, and Perl are widely used to write scripts that can automate the process of cleaning and preprocessing data. This may involve removing low-quality sequences, filtering out irrelevant data, or normalizing datasets for downstream analysis.

For example:

- In genomics, raw DNA sequences may contain errors due to sequencing technology limitations, and bioinformaticians write algorithms to correct these errors (error correction algorithms).
- In transcriptomics, RNA sequencing data is processed to quantify gene expression levels, which requires programming to handle the sequencing reads and align them to a reference genome.

Data Analysis and Algorithm Development

Programming is essential for designing and implementing algorithms that help analyze biological data. These algorithms can be used to solve a wide variety of problems, such as aligning DNA or protein sequences, identifying genes or regulatory elements, and predicting protein structures or interactions.

- **Sequence Alignment:** One of the most foundational tasks in bioinformatics is sequence alignment, where sequences of DNA, RNA, or proteins are compared to one another to find similarities or evolutionary relationships. Algorithms like BLAST (Basic Local Alignment Search Tool) or Needleman-Wunsch (global alignment) rely on programming for their implementation.

- **Gene Expression Analysis:** In transcriptomics, bioinformaticians use programming to analyze gene expression data (such as from RNA-Seq) to identify differentially expressed genes between different experimental conditions. Statistical programming in R or Python allows the implementation of various tools like DESeq2 or edgeR for this purpose.
- **Genome Assembly and Variant Calling:** When assembling genomes or calling variants (such as SNPs or indels), bioinformaticians rely heavily on programming to apply sophisticated algorithms that align sequencing reads to a reference genome, detect discrepancies, and predict genetic variants.

Data Visualization

Bioinformatics involves the interpretation of large datasets, and effective data visualization is essential for communicating results. Programming enables the creation of custom visualizations for complex biological data, allowing researchers to uncover patterns, trends, and outliers. Languages like R and Python, with libraries like ggplot2, matplotlib, and seaborn, are commonly used to generate high-quality visualizations such as heatmaps, phylogenetic trees, gene expression plots, and protein structures. Visual representations of biological networks or pathways can provide valuable insights into the functional relationships between genes or proteins and how they contribute to cellular processes or diseases.

Integration of Multidimensional Data

Biological data often comes from diverse sources (genomic, proteomic, metabolomic, etc.), and integrating these different data types to gain a holistic understanding of biological systems is a significant challenge in bioinformatics. Programming enables the development of frameworks and software that can integrate multiple layers of biological data, which is essential in systems biology and personalized medicine.

For example:

- **Multi-Omics Analysis:** By integrating genomic, transcriptomic, proteomic, and metabolomic data, bioinformaticians can build more accurate models of biological processes and disease states. Programming allows for the creation of pipelines that can seamlessly combine these datasets and generate meaningful insights.

Automation of Repetitive Tasks

Biological research often involves performing the same kinds of analyses on multiple datasets or performing repetitive tasks, such as data cleaning, analysis, and reporting. Programming allows researchers to automate these tasks, saving time and reducing human error. This automation is crucial for handling the large-scale data generated by high-throughput technologies.

- **Workflow Automation**: Bioinformaticians use tools like Snakemake or Nextflow to create reproducible, automated workflows that can process biological data from raw reads to final results in a systematic and consistent manner.

Development of Bioinformatics Tools and Software

The bioinformatics community relies on the development of specialized software tools and databases that can perform complex analyses, store biological data, and facilitate research. Programming is fundamental to creating and improving these tools, many of which are open-source and shared across the research community. Examples include genome browsers (like UCSC Genome Browser), structural bioinformatics tools (such as PyMOL for visualizing protein structures), or databases for storing genomic variants (like dbSNP).

Statistical Analysis and Machine Learning

Programming is increasingly important in bioinformatics for statistical analysis and machine learning (ML), which are used to make sense of large and complex biological datasets. Researchers use programming languages like R, Python, and Julia, along with specialized libraries (e.g., scikit-learn, TensorFlow, or caret), to apply statistical models and machine learning techniques to predict disease outcomes, identify biomarkers, or understand genetic interactions.

For example:

- **Cancer Genomics**: Machine learning algorithms can be used to identify genetic mutations that are associated with different cancer types, predict the prognosis of patients, and even help in the development of personalized treatments.
- **Pharmacogenomics**: Machine learning can be used to predict how individual patients will respond to certain drugs based on their genetic makeup.

Computational Biology and Modeling

Bioinformatics often involves the creation of computational models to simulate biological processes. These models may include simulations of protein folding, molecular docking, or biochemical pathways. Programming languages like Python, C++, or Julia are commonly used to develop these models, and specialized libraries (e.g., BioPython, BioPerl) help streamline the process.

- **Protein Structure Prediction:** Using algorithms such as Rosetta or AlphaFold, programming plays a key role in predicting the 3D structure of proteins from their amino acid sequences, which is crucial for drug discovery and understanding disease mechanisms.
- **Molecular Dynamics:** Programming is essential in computational chemistry and molecular dynamics simulations to model the behavior of molecules and understand their interactions.

Programming is the cornerstone of bioinformatics. From data collection and preprocessing to advanced statistical analysis and the development of novel algorithms and tools, programming enables bioinformaticians to extract valuable insights from biological data. As the field continues to evolve, the integration of advanced programming techniques with other disciplines like machine learning and artificial intelligence will continue to drive innovation in bioinformatics, revolutionizing fields such as genomics, personalized medicine, and drug discovery. With its ability to handle big data, automate processes, and facilitate the development of new algorithms and tools, programming remains indispensable to modern biological research.

BASIC COMPUTATIONAL THINKING

Computational thinking (CT) is a problem-solving process that involves breaking down complex tasks into manageable parts, identifying patterns, formulating step-by-step instructions, and creating solutions that can be executed by a computer or other automated systems. At its core, computational thinking is not limited to computer science but is a fundamental way of thinking that can be applied across various disciplines to address problems methodically and efficiently. It involves key concepts like abstraction, decomposition, pattern recognition, and algorithm design, which enable individuals to approach problems in a logical and structured way. As we increasingly rely on digital technologies and data-driven decision-making in our everyday lives, computational thinking has become an essential skill in almost every field, from biology and engineering to economics and social sciences. It empowers individuals to break down challenges into simpler, more manageable steps, enabling solutions to be developed and optimized through systematic processes. In the context of education, business, or research, computational thinking fosters creativity and innovation by encouraging individuals to think critically and analytically about problems, and it provides the foundation for developing automated systems that can handle large, complex datasets or repetitive tasks. Whether you're programming a computer, designing a workflow, or simply trying to solve a real-world problem, computational thinking provides the framework for crafting effective and efficient solutions.

Computational thinking (CT) is the process of approaching problems in a way that can be solved through computational means, involving the use of logical reasoning and algorithms. It is a problem-solving framework that applies across a wide range of fields, from computer science to biology, engineering, economics, and even the arts. It doesn't require one to be an expert in programming or computer science, but it does involve thinking in ways that allow for the creation of efficient solutions to problems.

CT encourages breaking complex issues into simpler, more manageable steps and then applying systematic methods to solve those issues. The key principles of computational thinking – **decomposition, pattern recognition, abstraction, and algorithm design** – work together to help individuals think logically and methodically about the problems they encounter.

Decomposition

Decomposition is the process of breaking down a complex problem or system into smaller, more manageable parts. Rather than attempting to solve a large, overwhelming problem all at once, you divide it into smaller sub-problems that are easier to understand and tackle.

For example, consider the task of developing an online shopping platform. Instead of trying to design the entire system at once, you can decompose the task into smaller components, such as:

- User authentication (login and signup)
- Product catalog
- Shopping cart management
- Payment processing
- Order tracking

Each of these sub-problems can then be solved individually and in parallel, and their solutions can later be integrated into a unified system. This process of decomposition is vital in managing the complexity of real-world problems and ensuring that solutions are both practical and scalable.

Pattern Recognition

Pattern recognition involves identifying similarities or trends in the data or problem that can simplify the solution process. It is about noticing recurring elements that can lead to generalized rules or patterns, which can then be used to solve similar problems more efficiently in the future.

For example, when analyzing a large dataset of customer purchases, one might notice that people tend to buy certain products together (e.g., shoes and socks, or laptops and laptop bags). Recognizing these patterns can lead to creating recommendation systems or targeted marketing strategies that optimize customer experience and business performance.

Pattern recognition can be applied across various domains, such as:

- **Data analysis:** Recognizing trends in large datasets (e.g., in financial markets or climate science) can lead to insights and predictions.
- **Computer vision:** Identifying patterns in image data allows facial recognition software to work effectively.
- **Text processing:** Detecting patterns in language enables applications like spam detection or natural language processing (NLP).

In essence, pattern recognition helps to streamline problem-solving by leveraging known patterns to predict outcomes or automate processes.

Abstraction

Abstraction is the process of simplifying complex problems by focusing on the essential features while ignoring the irrelevant details. It allows you to generalize the solution so that it applies to a wide range of similar problems. In computational thinking, abstraction involves creating models or representations of the problem that capture only the most important aspects.

For instance:

- When building a software application, the user interface (UI) design can be abstracted from the underlying code so that developers can focus on the logic of the program without getting distracted by the specifics of how it looks.
- In the context of programming, abstraction allows us to work with high-level languages (like Python or Java) instead of dealing directly with low-level machine code. This enables faster development and easier troubleshooting.

Abstraction is also a cornerstone of **modularization** in programming, where complex programs are divided into smaller modules or functions that can be reused and maintained independently. By abstracting unnecessary details, you can create scalable and flexible systems that can be adapted to new requirements or environments without significant rework.

Algorithm Design

At the heart of computational thinking is the creation of **algorithms** – step-by-step instructions for solving a problem or completing a task. An algorithm is a well-defined procedure that takes an input, processes it in a defined way, and produces an output. Developing efficient algorithms is central to making computational solutions work quickly and accurately.

For example:

- A simple algorithm for sorting a list of numbers might start by comparing the first two numbers, swapping them if necessary, and then moving to the next pair of numbers, repeating the process until the list is sorted. More sophisticated algorithms, such as QuickSort or MergeSort, achieve the same result but with better performance, especially with large datasets.

Algorithms are used everywhere – from search engines, social media platforms, and video games to biomedical research and financial analysis. They can be designed to optimize tasks such as data retrieval, computation, resource allocation, or decision-making.

Automation and Efficiency

Once you've developed a method to solve a problem through decomposition, pattern recognition, abstraction, and algorithm design, the next step is often to automate the process to make it more efficient. Automation means creating systems or software that can carry out repetitive tasks without human intervention.

For example:

- A recommendation engine on a streaming platform like Netflix or Spotify uses algorithms to automatically suggest content based on users' past behaviors, such as viewing history or genre preferences.
- In a manufacturing system, robots might be programmed to perform assembly tasks using pre-determined algorithms, significantly speeding up production processes while reducing human error.

By automating repetitive tasks, computational thinking helps to save time and resources, improving both productivity and accuracy.

Debugging and Refinement

A critical part of computational thinking is debugging – the process of identifying and fixing errors or inefficiencies in an algorithm, program, or model. No solution is perfect at first, and refining your approach through testing, validation, and modification is an essential part of computational thinking. Debugging helps ensure that the solution not only works but works optimally under various conditions.

Real-World Applications of Computational Thinking

Computational thinking can be applied across a wide variety of fields, beyond just computer science. Some examples include:

Medicine: In bioinformatics, researchers use computational thinking to analyze genomic data, model biological systems, or predict patient outcomes based on genetic information.

- **Education:** Teachers can use CT to design personalized learning experiences by breaking down complex concepts into digestible chunks, identifying student learning patterns, and creating algorithms to recommend tailored resources.
- **Business:** Companies use computational thinking for market analysis, data-driven decision-making, and automating customer service tasks (such as chatbots or recommendation engines).
- **Social Sciences:** Economists and sociologists use CT to model societal systems, predict behaviors, and optimize resource allocation in areas such as urban planning or environmental conservation.
- **Robotics and AI:** Computational thinking is key in designing robots and artificial intelligence systems that can solve complex tasks by breaking them down into smaller parts, recognizing patterns, and learning from experience.

Computational thinking is a powerful, universal framework that provides structured, efficient approaches to problem-solving across various disciplines. Whether you're programming a computer, designing a workflow, or solving a problem in everyday life, the principles of decomposition, pattern recognition, abstraction, and algorithm design provide the foundation for crafting effective, scalable, and optimized solutions. As digital technologies continue to permeate every aspect of society, computational thinking will only become more important, enabling individuals to solve problems creatively and effectively in an increasingly complex world.

5 Chapter

BIOLOGICAL DATABASES

INTRODUCTION

Biological databases are indispensable tools in modern biology, supporting the organization, retrieval, and analysis of vast amounts of biological data. With the explosion of biological data, particularly in genomics, proteomics, and systems biology, these databases have become central to a wide range of research fields, from medicine and drug discovery to evolutionary studies and environmental science. They provide a structured and accessible framework for storing complex data on DNA sequences, protein structures, gene expression profiles, metabolic pathways, and much more. By organizing biological information in a systematic way, these databases not only allow researchers to access and compare datasets efficiently but also facilitate data mining, hypothesis generation, and the discovery of novel biological insights.

The development of biological databases has paralleled advancements in high-throughput technologies such as next-generation sequencing (NGS), microarrays, and mass spectrometry, which have enabled researchers to generate data at unprecedented scales. For instance, the sequencing of genomes has led to the creation of reference databases like GenBank and Ensembl, which house millions of nucleotide sequences and annotations for a wide variety of species. Similarly, protein-focused databases such as UniProt and Protein Data Bank (PDB) provide detailed information about protein sequences, structures, functions, and interactions, all of which are critical for understanding the molecular mechanisms of life. The integration of data from multiple biological layers, such as genomics, transcriptomics, and proteomics, within databases enables a holistic view of biological processes and their regulation, contributing to fields like systems biology and personalized medicine.

Biological databases also serve as platforms for collaboration and data sharing across research communities, fostering innovation through open access and standardization of biological data. Publicly accessible databases, such as the National Center for Biotechnology Information (NCBI) and European Bioinformatics Institute (EBI), allow scientists from around the world to contribute, curate, and explore data in a collaborative environment. This interconnectedness not only accelerates discoveries but also facilitates reproducibility and transparency in scientific research. As biological data continues to grow in volume and complexity, biological databases are evolving with new technologies like machine learning and artificial intelligence to enhance data analysis, prediction, and visualization. The future of biological databases lies in their ability to integrate diverse data types, manage large-scale datasets efficiently, and support data-driven discoveriess that have the potential to revolutionize healthcare, agriculture, and environmental science.

Biological databases and bioinformatics are at the heart of modern biological research and the burgeoning fields of genomics, systems biology, and personalized medicine. The rapid expansion of biological data, fueled by advances in high-throughput technologies such as next-generation sequencing (NGS), microarrays, and mass spectrometry, has necessitated the development of robust computational tools and structured databases to store, manage, analyze, and interpret these vast quantities of information. Biological databases house this data in an organized and accessible format, enabling researchers to make sense of complex biological systems, identify patterns, and generate new hypotheses that lead to groundbreaking discoveries.

Biological Databases

Biological databases are organized repositories that store data related to the molecular biology of living organisms, including DNA sequences, RNA sequences, protein sequences, structural data, gene expression profiles, metabolic pathways, and functional annotations. The role of biological databases is to provide a centralized resource where data can be stored, accessed, queried, and analyzed. These databases serve as a backbone for a variety of scientific disciplines such as genomics, proteomics, transcriptomics, and evolutionary biology. They have grown in importance as biological research has shifted towards high-throughput, large-scale data generation and integrative analysis.

Types of Biological Databases

Sequence Databases

Sequence databases are among the most fundamental and widely used resources in bioinformatics. These databases store vast amounts of information related to the sequences of nucleic acids (DNA and RNA) and proteins, providing critical data for a variety of biological applications. They serve as the backbone for many core research tasks, including sequence alignment, gene annotation, comparative genomics, and functional genomics. The information stored in sequence databases is essential for identifying genes, studying evolutionary relationships, discovering genetic variations, and understanding molecular functions. In DNA and RNA sequence databases, the stored information includes the exact nucleotide sequences of genes, exons, introns, regulatory regions, and untranslated regions (UTRs), along with metadata such as organism information, gene annotations, and functional predictions. Similarly, protein sequence databases house the amino acid sequences of proteins, which can be linked to information about protein function, structure, interactions, and pathways. One of the primary functions of sequence databases is to facilitate sequence alignment, a process in which researchers compare two or more biological sequences to identify regions of similarity. This is crucial for identifying conserved sequences, studying functional elements of genes, and detecting mutations or variants associated with diseases. By aligning sequences from different organisms, researchers can also infer evolutionary relationships and construct phylogenetic trees, which help trace the evolutionary history of species or specific genes. In comparative genomics, sequence databases allow for large-scale comparisons of genomes from multiple species, revealing insights into genetic diversity, conserved pathways, and species-specific adaptations. Moreover, sequence databases play a pivotal role in gene annotation, which involves predicting the locations of genes within a genome and assigning functional information to these genes. Gene annotation helps researchers understand the roles of individual genes, their regulatory mechanisms, and their involvement in biological processes. The combination of experimental data and computational tools allows for the generation of high-quality gene annotations, which are invaluable for understanding disease mechanisms and discovering novel therapeutic targets. By organizing and providing easy access to vast volumes of sequence data, these databases empower researchers to explore the molecular underpinnings of life at unprecedented scales. Whether it's unraveling the genetic basis of a disease, discovering new drug targets, or exploring the biodiversity of life on Earth, sequence databases are indispensable in modern biological research, driving both basic science and applied biotechnology forward. Some well-known sequence databases include:

- **GenBank:** A comprehensive public sequence database hosted by the National Institutes of Health (NIH), which contains nucleotide sequences and associated annotations for a variety of species.
- **UniProt:** A database of protein sequences and functional annotations, including detailed information on protein families, domains, and structures.
- **RefSeq:** A curated collection of reference sequences for genomes, transcriptomes, and proteomes, which helps in the accurate identification and functional annotation of genes.

Structural Databases

Structural databases are essential resources that store detailed information about the three-dimensional (3D) structures of biomolecules, particularly proteins, nucleic acids, and their complexes. The 3D structure of a biomolecule is intimately tied to its function, and understanding this structure is crucial for gaining insights into how biomolecules interact, how they carry out their biological roles, and how mutations may lead to disease. These databases are indispensable for researchers studying protein folding, protein-ligand interactions, macromolecular complexes, and structural genomics, as well as for those involved in drug design and systems biology. Proteins, for example, are linear chains of amino acids that fold into specific 3D structures, and these structures dictate their biological function. A single mutation in a protein's sequence can drastically alter its structure, potentially leading to loss of function or malfunction, which can contribute to diseases such as Alzheimer's, cystic fibrosis, or cancer. Structural databases provide the high-resolution 3D models of proteins, often determined through experimental techniques like X-ray crystallography, nuclear magnetic resonance (NMR) spectroscopy, or cryo-electron microscopy (cryo-EM). By storing this structural data, these databases enable researchers to visualize how proteins fold into their functional conformations and understand the structural basis of their activity. One of the most widely known and utilized structural databases is the Protein Data Bank (PDB), which houses an ever-growing collection of 3D protein structures. The PDB allows scientists to access not only the atomic coordinates of a molecule but also its secondary and tertiary structural features, such as alpha-helices, beta-sheets, and binding sites. Researchers can download these structural models, analyze them with computational tools, and simulate molecular dynamics to investigate how proteins behave in different environments. The PDB also integrates metadata, such as experimental conditions, methods used to solve the structure, and details about molecular interactions, making it a comprehensive resource for structural biology. In addition to protein structures, structural databases may also contain information on the 3D structures of nucleic acids (DNA, RNA), protein-RNA complexes, and other biomolecular assemblies. For example, databases like RCSB PDB, Nucleic Acid Database (NDB), and RNAcentral provide structures and sequences for nucleic acid-based structures, which are important for understanding the functional roles of RNA in gene regulation, protein synthesis, and more. Structural annotations within these databases are critical for researchers seeking to understand how specific molecular features influence function. For instance, researchers can use structural databases to identify active sites in enzymes, binding pockets for drug design, and regions that are conserved across species, suggesting their importance in biological processes. By leveraging structural data, scientists can also predict the effects of mutations on protein folding and function, providing insights into genetic diseases or offering strategies for designing therapeutic molecules. Furthermore, structural databases play a key role in drug discovery and molecular docking. In silico techniques like molecular docking rely on detailed 3D structures to predict how small molecules, including potential drugs, interact with target proteins. The information stored in structural databases allows drug developers to identify potential drug binding sites, model the interactions between drug candidates and their targets, and refine compounds for better specificity and efficacy. The availability of detailed, high-resolution 3D models accelerates the process of drug discovery by providing a platform for virtual screening, reducing the need for costly and time-consuming experimental work in the early stages of development. Overall, structural databases are indispensable for advancing our understanding of the molecular mechanisms that drive life. By providing a comprehensive, accessible archive of 3D molecular structures, these databases empower researchers to explore the molecular basis of disease, discover new therapeutic targets, and design drugs with greater precision and efficacy. As technologies in structural biology continue to advance, the role of structural databases in biological research will only grow, unlocking new frontiers in medicine, biotechnology, and molecular biology. Structural databases are crucial for understanding protein folding, function, and interactions. They also support drug discovery by providing insights into potential drug targets. Prominent structural databases include:

- **Protein Data Bank (PDB):** A repository of 3D structural data of proteins, nucleic acids, and other macromolecules, which are determined through experimental techniques like X-ray crystallography and NMR spectroscopy.
- **SCOP (Structural Classification of Proteins):** A database that classifies proteins based on their structural and evolutionary relationships.

Functional and Pathway Databases

Functional and pathway databases are essential resources in modern biological research, providing deep insights into the roles of genes, proteins, and other biomolecules within the context of cellular processes, metabolic networks, and complex signaling pathways. These databases are invaluable for understanding the biological functions of molecular components, how they interact with one another, and how these interactions orchestrate broader physiological and pathological states. By mapping out how genes and proteins collaborate within metabolic pathways, signal transduction networks, and gene regulation circuits, functional and pathway databases serve as a critical framework for researchers seeking to understand the molecular basis of health and disease. These databases go far beyond storing simple sequence information; they integrate diverse molecular data, linking genetic sequences with their functional roles in cellular systems.

They provide essential annotations on gene functions, protein activities, and biological processes, allowing researchers to predict how changes at the genetic level may impact cellular behavior, metabolic functions, and signal transduction. This functionality is particularly important for researchers working in systems biology, where understanding the holistic and interconnected nature of molecular networks is key to deciphering the complexities of biological systems.

- **Gene and Protein Function Annotation:** Functional databases enable the classification and annotation of genes and proteins based on their **biological roles.** For instance, databases like **Gene Ontology (GO)** offer a controlled vocabulary for describing gene functions across three main categories: **biological processes** (e.g., cell division, DNA repair), **molecular functions** (e.g., enzyme activity, ligand binding), and **cellular components** (e.g., mitochondrion, cytoskeleton). This standardized system allows for the consistent description of gene and protein functions across different organisms, promoting comparisons and facilitating the identification of conserved biological mechanisms.
- **Metabolic Pathways:** Metabolic pathways are intricate biochemical networks that govern the synthesis, breakdown, and transformation of biomolecules, such as carbohydrates, lipids, and amino acids, to maintain cellular homeostasis. Pathway databases like **KEGG (Kyoto Encyclopedia of Genes and Genomes)**, **Reactome**, and **BioCyc** map these metabolic pathways, linking genes to specific enzymes and metabolites involved in cellular metabolism. This is crucial for understanding how cells generate energy, synthesize essential molecules, and regulate biosynthetic pathways in response to environmental cues. By visualizing these metabolic networks, researchers can identify key regulatory enzymes, metabolic bottlenecks, and points of therapeutic intervention in diseases like diabetes, cancer, and metabolic syndromes.
- **Signaling Pathways and Cell Communication:** Signaling pathways are a core component of cellular communication, enabling cells to respond to external stimuli like hormones, growth factors, stress signals, and nutrients. Functional and pathway databases are essential for mapping these **signal transduction networks**, which often involve cascades of molecular interactions, including **receptors**, **kinases**, **transcription factors**, and **second messengers**. Databases such as **Reactome**, **SIGNOR**, and **WikiPathways** provide detailed information on how signals are transmitted across the cell membrane, how intracellular pathways are activated or repressed, and how gene expression is regulated in response to signals. Understanding these pathways is critical for uncovering how

cells regulate growth, differentiation, apoptosis, and immune responses, as well as for understanding the molecular basis of diseases like cancer, autoimmune disorders, and neurodegenerative diseases.

- **Pathogen Interaction and Immune Response**: Many pathway databases also focus on the molecular interactions between pathogens (such as bacteria, viruses, and parasites) and their host organisms. These interactions can significantly alter the normal functioning of host cellular pathways, promoting infection, immune evasion, and disease progression. For example, **Reactome** and **KEGG** provide pathway data on how pathogens manipulate host cell machinery, hijack signaling pathways, or disrupt immune responses. Additionally, these databases help researchers explore the host's immune system pathways, including the activation of **T cell** signaling, **cytokine release**, and the regulation of immune checkpoints. This is crucial for developing vaccines, immunotherapies, and targeted interventions for infectious diseases, as well as for better understanding immune-related conditions such as autoimmune diseases and immunodeficiencies.
- **Disease Pathways and Therapeutic Targeting**: A key application of functional and pathway databases is in **disease research**. Many diseases, including cancer, cardiovascular diseases, neurodegenerative disorders, and metabolic diseases, arise from dysregulation of critical molecular pathways. By exploring how specific mutations or alterations in pathway components contribute to disease, researchers can identify potential therapeutic targets. For example, the **PI3K-AKT pathway**, which regulates cell growth and survival, is often upregulated in various cancers. Pathway databases allow researchers to track the molecular details of this pathway, including the genes, proteins, and small molecules involved, helping to identify new druggable targets or biomarkers for early disease detection. Understanding disease pathways also helps scientists identify how treatments, such as small molecule inhibitors or biologics, can interfere with these dysregulated pathways.
- **Cross-Species Pathway Comparison and Evolutionary Insights:** Functional and pathway databases also facilitate cross-species comparisons, allowing researchers to study the evolutionary conservation of pathways across different organisms. By examining conserved pathways in model organisms like yeast, fruit flies, or mice, researchers can gain insights into fundamental biological processes that are shared across species. These comparisons can highlight important pathways that are evolutionarily preserved due to their critical roles in cellular function and survival. For

example, the comparison of metabolic or signaling pathways between humans and model organisms can reveal conserved mechanisms of gene regulation or protein interactions, providing valuable insights into basic cellular functions and disease biology.

- **Integrating Multi-Omics Data for Systems Biology:** In addition to linking genes and proteins to specific pathways, modern functional and pathway databases integrate data from various "omics" layers—such as genomics, transcriptomics, proteomics, and metabolomics—to offer a holistic view of biological processes. This **multi-omics** approach allows researchers to understand how changes at the genetic level (mutations, gene expression) affect the proteome (protein expression, modifications), which in turn influences metabolic networks and signaling pathways. By integrating multi-omics data, these databases enable more accurate models of biological systems, which are crucial for advancing **systems biology**, studying gene-environment interactions, and developing personalized medicine approaches.

Prominent Functional and Pathway Databases

- **KEGG (Kyoto Encyclopedia of Genes and Genomes):** KEGG is one of the most comprehensive pathway databases, offering detailed maps of metabolic, signaling, and disease pathways. It links genes and proteins to specific enzymatic functions, cellular processes, and diseases, providing an integrated view of biological functions at the molecular level.
- **Reactome:** Reactome provides an open-source, curated pathway database focused on human biology. It offers an extensive catalog of molecular interactions and reactions, from metabolic pathways to signaling cascades, and it is widely used for pathway visualization, analysis, and interpretation in biomedical research.
- **WikiPathways:** WikiPathways is a community-driven, open-access database of biological pathways. It allows researchers to explore both well-established and newly discovered pathways, as well as contribute and update content. This collaborative approach ensures that the database remains dynamic and current, incorporating the latest findings in pathway research.
- **BioCyc:** BioCyc is a collection of pathway and genome databases that provide detailed metabolic network models for thousands of organisms, from bacteria to plants and humans. It is particularly useful for studying microbial metabolism and environmental interactions.

- **SIGNOR:** SIGNOR is a pathway database focused on signal transduction networks. It provides curated data on molecular interactions involved in cell signaling, with an emphasis on human biology. SIGNOR is an invaluable tool for researchers studying cellular responses to stimuli, as well as for understanding the molecular underpinnings of diseases like cancer and diabetes.

Functional and pathway databases are integral to our understanding of the molecular mechanisms that govern life. By linking genes, proteins, and small molecules to their roles in metabolic and signaling pathways, these databases provide invaluable insights into cellular processes, disease mechanisms, and potential therapeutic targets. Whether used to explore gene function, map out disease pathways, or design new drugs, functional and pathway databases are essential resources in the fields of **bioinformatics**, **systems biology**, **drug discovery**, and **precision medicine**. As research continues to generate vast amounts of molecular data, these databases will play an increasingly important role in translating complex biological information into actionable knowledge, ultimately paving the way for more effective treatments and better health outcomes.

Gene Expression Databases

These databases store data related to gene expression patterns under various conditions, including differential expression, tissue-specific expression, and developmental stages. Examples include:

- **GEO (Gene Expression Omnibus):** A repository for high-throughput gene expression data and series of experiments, allowing researchers to explore transcriptional profiles.
- **ArrayExpress:** A database that houses gene expression data from microarray and RNA-Seq experiments, providing tools for gene expression analysis.

Variant Databases

Variant databases are critical resources that focus on cataloging and characterizing genetic variations within populations, offering essential insights into how these variations contribute to **human health, disease susceptibility**, and **evolutionary processes.** Genetic variation is a natural and fundamental aspect of biology, and these databases track various forms of genetic alterations, including **single nucleotide polymorphisms (SNPs), insertions, deletions (indels),** and **structural variations (SVs).** By cataloging these genetic changes, variant databases provide a comprehensive resource for understanding the genetic basis of **complex traits, diseases**, and **phenotypic diversity** across populations.

- **Types of Genetic Variants Cataloged in Variant Databases**
- **Single Nucleotide Polymorphisms (SNPs):**SNPs are the most common form of genetic variation, involving a change in a single nucleotide base pair in the DNA sequence. Although most SNPs have no effect on gene function, some can influence gene expression or protein function, and certain SNPs are associated with an increased risk of diseases, including cancer, cardiovascular diseases, and neurodegenerative disorders. Variant databases track the location of SNPs within the genome, along with information about their allele frequency, associated phenotypic effects, and any links to specific diseases. Notable databases like **dbSNP** provide a rich catalog of SNPs, enabling researchers to study how these variations correlate with health outcomes and contribute to personalized medicine.
- **Insertions and Deletions (Indels):** Indels refer to the insertion or deletion of small segments of DNA, which can range from a single nucleotide to several thousand base pairs. Indels can disrupt the reading frame of a gene or affect regulatory regions, potentially leading to changes in protein function or gene expression. These variations are particularly important in studying genetic diseases, as some indels are associated with conditions like **cystic fibrosis**, **Duchenne muscular dystrophy**, and **sickle cell anemia**. Variant databases like **gnomAD** and **ClinVar** offer detailed annotations of indels, highlighting their potential pathogenicity and clinical significance, making them invaluable tools for clinical genetics and genetic counseling.
- **Structural Variations (SVs):** Structural variations refer to large-scale alterations in the genome, including **copy number variations (CNVs)**, **inversions**, **translocations**, and **large deletions or duplications**. These variations can affect gene dosage, disrupt important genes, or rearrange the genome in ways that contribute to complex diseases such as **schizophrenia**, **autism spectrum disorders**, and **cancer**. SVs are often challenging to detect due to their size, but advancements in sequencing technologies have enabled more comprehensive detection and cataloging of structural variants. Variant databases like **DECIPHER** and **Database of Genomic Variants (DGV)** provide information on known SVs, their clinical relevance, and their potential role in genetic disorders.

Roles and Applications of Variant Databases

- **Disease Association and Genetic Epidemiology:** Variant databases are instrumental in identifying genetic variations associated with diseases, including rare and complex disorders. By aggregating information from population-based studies, case-control studies, and clinical genetic data, these databases help identify genetic risk factors for a wide range of conditions, including **cardiovascular diseases**, **autoimmune disorders**, and **neurodevelopmental conditions**. Researchers can perform association studies to link specific variants to disease phenotypes, paving the way for better diagnostic tools and targeted therapies.

- **Clinical Genetic Testing and Personalized Medicine**: Variant databases play a crucial role in **clinical genetics** by providing clinicians with curated information on genetic variants that have known clinical significance. For example, **ClinVar** is a widely used database that classifies genetic variants based on their pathogenicity (e.g., benign, likely pathogenic, pathogenic), helping healthcare providers interpret the results of genetic tests. As the field of **precision medicine** advances, these databases enable the identification of genetic variants that influence how patients respond to specific drugs or therapies. This information can be used to develop personalized treatment plans that are tailored to an individual's unique genetic makeup, improving treatment efficacy and minimizing adverse drug reactions.

Pharmacogenomics

Pharmacogenomics is the study of how genetic variation affects an individual's response to drugs. Variant databases are essential for identifying genetic variants that influence drug metabolism, efficacy, and toxicity. For instance, certain SNPs in genes like **CYP450** can affect how well a person metabolizes drugs, influencing drug choice and dosage. By cataloging pharmacogenomic variants, these databases help guide drug development and the clinical application of personalized medicine.

- **Population Genetics and Evolutionary Studies**: Variant databases are invaluable for studying **population genetics** and the evolutionary dynamics of genetic variation. By aggregating data from diverse populations, these databases allow researchers to investigate patterns of genetic diversity, evolutionary adaptations, and the history of human migration. For example, the **1000 Genomes Project** provides detailed information about genetic variation in human populations around the world, offering insights into human evolution, the genetic basis of adaptation to different environments, and the origin of complex diseases.
- **Variant Interpretation and Pathogenicity Prediction**: A major challenge in the study of genetic variation is determining the clinical significance of rare or novel variants. Variant databases often provide tools for **variant interpretation**, integrating data from multiple sources, including functional predictions, evolutionary conservation, population frequency, and clinical reports. These resources help researchers and clinicians assess

whether a variant is likely to be pathogenic or benign. Tools like **PolyPhen** and **SIFT** predict the impact of amino acid changes on protein structure and function, complementing the information in variant databases and aiding in the identification of disease-causing variants.

Prominent Variant Databases

- **dbSNP (Single Nucleotide Polymorphism Database):** dbSNP, hosted by the National Center for Biotechnology Information (NCBI), is one of the most widely used databases for cataloging SNPs. It provides information on millions of SNPs from various organisms, including humans. dbSNP allows researchers to access detailed annotations, including allele frequencies, associated diseases, and the functional impact of specific SNPs.
- **gnomAD (Genome Aggregation Database):** gnomAD aggregates data from large-scale sequencing projects and offers a comprehensive catalog of genetic variation across different populations. It is particularly valuable for understanding the population frequency of genetic variants, helping researchers identify rare variants that may contribute to disease.
- **ClinVar:** ClinVar is a public database that provides detailed information on the relationship between genetic variants and disease. It categorizes variants based on their clinical significance, such as pathogenic, likely pathogenic, benign, or of uncertain significance. ClinVar is widely used by clinicians and researchers for interpreting the results of genetic testing.

DECIPHER (Database of Chromosomal Imbalance and Phenotype in Humans Using Ensembl Resources)

DECIPHER is a database that specializes in structural variation (SV) and its relationship to human disease. It provides detailed clinical and genomic information on rare genetic disorders caused by structural variations, helping researchers understand the impact of SVs on phenotypic outcomes.

- **DGV (Database of Genomic Variants):** DGV catalogs structural variations, including copy number variations (CNVs), insertions, deletions, and inversions. It provides data on the frequency and distribution of these variants in diverse populations, offering valuable information for understanding their role in disease and evolution.
- **ExAC (Exome Aggregation Consortium):** ExAC, now integrated into gnomAD, is a large-scale resource for studying human genetic variation, particularly focusing on exonic variants (those found within coding regions of the genome). ExAC's data helps identify rare variants and their association with diseases, providing a comprehensive picture of exonic variation in humans.

Variant databases are fundamental to modern genetics and genomics, providing detailed information about the vast array of genetic variations that exist within populations. They play a crucial role in **disease research**, **clinical genetics**, **population studies**, and **personalized medicine.** By cataloging and annotating genetic variants, these databases enable researchers to identify disease-causing mutations, predict the effects of novel variants, and understand the genetic underpinnings of complex traits. As sequencing technologies continue to advance and the amount of genetic data grows, variant databases will remain essential tools for advancing our understanding of genetics and improving human health.

Specialized Databases

In addition to the broad-spectrum biological databases that encompass a wide range of biomolecules and species, **specialized databases** focus on specific types of biomolecules, cellular processes, or particular organisms. These niche databases provide in-depth information about particular aspects of biology, allowing researchers to dive deeply into specialized fields of study. Whether it›s a focus on **specific classes of molecules**, **organisms of interest**, or **biological phenomena**, specialized databases are invaluable tools for researchers seeking high-resolution insights that go beyond general-purpose databases.

Protein and Enzyme Databases: Protein and enzyme databases are dedicated to cataloging detailed information about protein sequences, structures, functions, and interactions. These resources are indispensable for researchers studying **protein function**, **enzyme mechanisms**, and **protein-protein interactions**.

- **UniProt** is one of the most well-known and comprehensive protein databases, providing detailed annotations on protein sequences, structures, functional domains, and post-translational modifications. It includes entries from multiple organisms and is often used for functional characterization, sequence alignment, and comparative genomics.
- **BRENDA** is a specialized enzyme database that provides detailed information on enzyme functions, kinetics, and substrate specificity. Researchers use BRENDA for enzyme classification, metabolic pathway analysis, and identifying potential targets for drug design.
- **Protein Data Bank (PDB)** focuses on storing three-dimensional structures of proteins and other biomolecules like nucleic acids. It provides a repository of experimentally determined structures and is used to explore molecular interactions and design structure-based drug therapies.

Gene Expression Databases: Gene expression databases specifically focus on providing insights into **how genes are regulated** across different conditions, tissues, and stages of development. These databases are essential for **transcriptomics** and

functional genomics research, allowing researchers to examine patterns of gene expression and identify genes involved in disease or developmental processes.

- **GEO (Gene Expression Omnibus)** is a public database hosted by the NCBI, offering gene expression data from a wide variety of organisms and experimental conditions. GEO serves as a key resource for understanding gene expression profiles in diseases, drug treatments, and environmental responses.
- **ArrayExpress** is another major gene expression repository that provides data from high-throughput experiments, such as **microarrays** and **RNA sequencing**. Researchers utilize ArrayExpress to study differential gene expression, identify biomarkers, and explore gene regulatory networks.
- **The Cancer Genome Atlas (TCGA)** is a specialized resource that focuses on the genetic characterization of cancer. TCGA integrates genomics, transcriptomics, and clinical data from various cancer types, offering detailed insights into the molecular drivers of cancer and potential therapeutic targets.

Pathogen-Specific Databases

For researchers studying infectious diseases, **pathogen-specific databases** focus on the genomic and functional characterization of bacteria, viruses, fungi, and parasites. These databases are crucial for understanding microbial virulence, evolution, and drug resistance mechanisms, and they support the development of diagnostics, vaccines, and therapeutic interventions.

- **Virus Pathogen Database and Analysis Resource (ViPR)** provides a comprehensive resource for the genomic analysis of viruses. It supports researchers working on viruses like **HIV**, **influenza**, **hepatitis**, and **coronaviruses**, offering information on viral sequences, structure, taxonomy, and clinical data.
- **The Plasmodium Genome Database** focuses specifically on the malaria-causing parasite **Plasmodium.** It provides genetic, genomic, and functional data to support research into malaria biology, vaccine development, and drug discovery.
- **TB Database** focuses on the **Mycobacterium tuberculosis** pathogen, which causes tuberculosis. This database integrates genomic, transcriptomic, and proteomic data related to **M. tuberculosis**, facilitating research on **antibiotic resistance, virulence factors**, and the **immune response** to TB.

Species-Specific Databases

Many specialized databases are dedicated to particular species or groups of organisms, providing comprehensive resources for studying their biology in great detail. These databases are crucial for evolutionary research, comparative genomics, and the study of model organisms.

- **ZFIN (Zebrafish Information Network)** is a comprehensive database dedicated to the model organism **zebrafish.** It includes information on **genetic sequences**, **mutants**, **phenotypic data**, and **developmental processes**, making it a valuable resource for researchers studying vertebrate development, disease modeling, and drug discovery.
- **FlyBase** focuses on the **fruit fly (Drosophila melanogaster)**, a key model organism in genetics and developmental biology. FlyBase contains detailed information on **genomic sequences**, **mutations**, **phenotypes**, and **gene expression patterns**, supporting a wide range of genetic and biological research.
- **WormBase** is a database dedicated to the nematode **Caenorhabditis elegans**, a model organism used in studies of development, aging, and neurobiology. WormBase includes comprehensive data on the **genome**, **genetic variants**, **transcriptomics**, and **functional genomics** of C. elegans, offering insights into basic biological processes.

Metabolomics Databases: Metabolomics databases specialize in cataloging **metabolites** and **metabolic pathways** within organisms. These databases are essential for understanding **metabolism**, **biochemical networks**, and how **small molecules** influence health and disease. They support research in areas such as **drug metabolism**, **toxicology**, and **nutrition.**

- **HMDB (Human Metabolome Database)** provides detailed information on human metabolites, including their chemical properties, biological functions, and associated diseases. HMDB is a valuable resource for researchers exploring **metabolic disorders**, **biomarker discovery**, and **nutritional sciences.**
- **KEGG** also includes **metabolic pathways** and **drug metabolism** information, linking **metabolites** with enzymes and genes involved in their biosynthesis and catabolism. This enables researchers to understand **disease-associated metabolic disruptions** and to design drugs that target specific metabolic pathways.
- **METLIN** is another comprehensive metabolomics resource, focusing on a wide variety of metabolites, including **lipids**, **amino acids**, and **secondary**

metabolites. METLIN provides detailed spectral data, aiding in **metabolite identification** and **quantitative analysis** for researchers in **biochemistry** and **drug development.**

Genome-Wide Association Study (GWAS) Databases

GWAS **databases** focus on cataloging genetic variants associated with specific traits, diseases, or phenotypes. These databases support researchers working in **genetic epidemiology** and **personalized medicine**, enabling the identification of **genetic risk factors** for complex diseases.

- **GWAS Catalog** is an essential resource that compiles results from genome-wide association studies, linking genetic variants to a wide range of traits and diseases. The GWAS Catalog allows researchers to identify potential genetic risk factors for diseases such as **diabetes**, **heart disease**, and **neurodegenerative disorders.**
- **UK Biobank** is a large-scale resource that links genetic data with health and lifestyle information from over 500,000 participants. It is a key resource for investigating the genetic underpinnings of common diseases and for discovering new therapeutic targets.

RNA Sequence Databases

These databases are dedicated to **RNA sequencing data** and **RNA modifications**, offering critical insights into gene regulation, alternative splicing, and **non-coding RNA** function. They are especially useful in **transcriptomics**, **RNA biology**, and **epitranscriptomics** research.

- **RNAcentral** is a comprehensive resource that provides access to **non-coding RNA sequences** across a wide range of organisms. It integrates data from multiple sources, enabling researchers to explore the roles of **microRNAs**, **long non-coding RNAs**, and other RNA species in gene regulation and disease.
- **ENCODE (Encyclopedia of DNA Elements)** is an ongoing project that catalogs functional elements in the human genome, including **transcription factors**, **enhancers**, and **splicing sites.** It provides valuable data on how RNA is transcribed and processed in various cell types.

Specialized databases are vital for providing deep, focused insights into specific areas of biological research. Whether studying proteins, genes, metabolites, or pathogens, these databases offer high-resolution information that general-purpose databases cannot match. By concentrating on specific biomolecules, organisms, or biological processes, specialized databases support researchers across a range of disciplines, from molecular biology to drug discovery and clinical research. These resources continue to evolve as our understanding of biological systems becomes more refined, and they play a pivotal role in advancing science, improving healthcare, and driving innovations in personalized medicine.

There are also specialized databases that focus on specific types of biomolecules or particular organisms. For example:

- **TCGA (The Cancer Genome Atlas):** A resource for cancer genomics that includes genomic, transcriptomic, and epigenomic data for various cancer types.
- **Metagenomics Databases:** These include databases like IMG (Integrated Microbial Genomes), which store genomic and environmental data for microbes and communities of organisms.

Bioinformatics: The Bridge Between Data and Discovery

Bioinformatics is the interdisciplinary field that combines biology, computer science, and statistics to analyze and interpret large-scale biological data. The primary objective of bioinformatics is to extract meaningful insights from biological datasets, often using computational tools and algorithms. It plays a key role in areas like genome sequencing, systems biology, drug discovery, and evolutionary studies.

Bioinformatics encompasses a broad range of tasks, including:

- **Data Mining and Pattern Recognition:** One of the fundamental tasks in bioinformatics is the identification of patterns within large biological datasets. This could involve searching for conserved genetic sequences across different species, identifying biomarkers for diseases, or discovering new protein-protein interactions. Advanced techniques in machine learning, statistical modeling, and clustering are often used for these tasks.
- **Sequence Alignment and Comparison:** Sequence alignment is a foundational tool in bioinformatics, used to compare DNA, RNA, or protein sequences to identify similarities, evolutionary relationships, and functional regions. Algorithms like BLAST (Basic Local Alignment Search Tool) are widely used for this purpose. Sequence alignment can also be used to identify mutations or variations that may be linked to disease.
- **Genomic Data Analysis:** With the advent of high-throughput sequencing technologies, bioinformaticians analyze large-scale genomic data to assemble genomes, annotate genes, and study variation at the genomic level. This involves using algorithms to map sequence data to a reference genome, detect mutations, and perform comparative genomics across species.

- **Structural Bioinformatics:** Structural bioinformatics focuses on the analysis and prediction of the 3D structures of proteins and other biomolecules. The goal is to understand how the structure of a molecule relates to its function and to predict the effects of mutations. Tools such as molecular docking, protein structure prediction algorithms (e.g., Rosetta), and molecular dynamics simulations are commonly used.
- **Systems Biology and Network Analysis:** Systems biology seeks to understand biological systems as a whole by integrating data from genomics, transcriptomics, proteomics, and metabolomics. Bioinformatics plays a critical role in this integration, helping to build models of biological networks (e.g., gene regulatory networks, protein interaction networks, metabolic pathways) and simulating their behavior under different conditions.
- **Comparative Genomics and Evolutionary Biology:** Bioinformatics tools allow scientists to compare the genomes of different species, uncovering evolutionary relationships and the genetic basis of traits. Phylogenetic analysis, which uses sequence data to construct evolutionary trees, is an important application of bioinformatics in evolutionary biology.
- **Drug Discovery and Personalized Medicine:** Bioinformatics is revolutionizing drug discovery by enabling the identification of potential drug targets, screening of compounds, and prediction of drug interactions. In personalized medicine, bioinformatics tools are used to analyze patient-specific genomic data, helping clinicians to tailor treatments based on individual genetic profiles.
- **Visualization:** Given the complexity of biological data, bioinformatics tools also provide sophisticated visualization methods, such as heatmaps, gene expression plots, protein structure diagrams, and network graphs, which make it easier to interpret results and share findings with the scientific community.

Integration of Biological Databases with Bioinformatics

The power of biological databases is greatly enhanced when combined with bioinformatics tools. These databases provide the raw data, while bioinformatics techniques offer the methods for analyzing, interpreting, and making sense of that data. For example, a bioinformatician may use a genomic database like GenBank to obtain DNA sequences and then apply sequence alignment algorithms to compare them across species or to identify mutations in a disease-related gene. Similarly, protein structure databases such as the Protein Data Bank (PDB) can be combined with molecular dynamics simulations to predict how a protein will behave under different conditions or interact with potential drug molecules. Moreover, the integration of multiple types of data (genomics, transcriptomics, proteomics, etc.) within bioinformatics pipelines is a growing trend, particularly in systems biology and precision medicine. This allows for a more comprehensive understanding of biological processes and the development of therapeutic strategies tailored to individual patients' genetic profiles.

FUTURE DIRECTIONS

As both biological databases and bioinformatics continue to evolve, there are several key trends shaping their future. The increasing integration of artificial intelligence (AI) and machine learning (ML) into bioinformatics is expected to accelerate the analysis of complex biological datasets, enabling researchers to uncover hidden patterns and make predictions with greater accuracy. Additionally, the growing availability of single-cell RNA sequencing data, as well as advancements in spatial transcriptomics, is driving the need for new computational tools and databases that can handle these high-dimensional, multi-modal datasets. The continuous expansion of biological databases—both in terms of the number of organisms and the richness of the data—will further fuel discoveries in areas like drug discovery, gene therapy, and environmental genomics. Collaborative, open-access platforms will become more important as the scientific community moves toward a more global, data-sharing ecosystem. In conclusion, biological databases and bioinformatics will continue to be foundational to the future of biological research, medicine, and biotechnology, enabling us to better understand the complexities of life and harness that knowledge to improve human health and well-being.

PUBLIC DATABASES (GENBANK, EMBL, UNIPROT, ETC.)

Public databases like GenBank, EMBL, and UniProt are foundational pillars of modern biological and biomedical research. These open-access repositories provide researchers worldwide with vast, curated collections of genetic, genomic, and proteomic data, fueling discoveries across multiple disciplines, from genomics and evolutionary biology to drug development and personalized medicine. By offering freely available, up-to-date information on DNA sequences, protein structures, genetic variation, and functional annotations, these public databases enable the scientific community to share knowledge, collaborate on global research initiatives, and accelerate scientific advancements. The democratization of biological data through these databases has revolutionized our ability to explore the molecular mechanisms underlying health, disease, and life itself, ensuring that research findings are accessible to all and fostering a more inclusive, transparent scientific environment.

PUBLIC DATABASES: THE CORNERSTONE OF BIOLOGICAL DATA SHARING AND DISCOVERY

Public biological databases, such as **GenBank**, **EMBL**, **UniProt**, and others, are essential resources that support scientific discovery, data sharing, and collaboration across the globe. These repositories host vast amounts of biological information, ranging from genomic sequences and protein structures to functional annotations and genetic variation. By providing free access to these datasets, public databases have transformed biological research, enabling researchers to explore fundamental questions in molecular biology, comparative genomics, evolution, disease mechanisms, and drug discovery. The accessibility and extensive coverage of these databases make them indispensable tools in genomics, bioinformatics, systems biology, and translational medicine.

GenBank: The Global DNA Sequence Repository

GenBank, maintained by the **National Center for Biotechnology Information (NCBI)**, is one of the most widely used public databases for nucleotide sequences. Since its inception in 1982, it has grown into a comprehensive and global resource for storing and sharing **DNA and RNA sequences** from diverse organisms, including bacteria, viruses, plants, animals, and humans. GenBank serves as the central hub for sequence data submission, allowing researchers to deposit newly sequenced genomes, genes, and other genomic elements for free public access. GenBank is not just a sequence repository; it also provides associated annotations such as **gene function**, **genomic location**, and **variant information**. By providing an open-access platform for genomic data, GenBank enables scientists to perform a variety of analyses, including **sequence alignment**, **phylogenetic studies**, and **comparative genomics**. It is integrated with other NCBI resources, such as **BLAST** (Basic Local Alignment Search Tool), allowing users to rapidly compare their sequences against a vast database of known sequences to find similarities and infer functional and evolutionary relationships.

Features:

- **Public Access:** GenBank offers free access to the latest genomic sequences and annotations.
- **Regular Updates:** The database is updated daily with new data submitted by researchers worldwide.
- **Global Coverage:** GenBank hosts sequences from virtually all organisms, from microbes to humans.

Applications:

- **Genomic Research:** Analyzing genetic diversity, studying the molecular basis of diseases, and identifying novel genes.

- **Comparative Genomics:** Identifying conserved sequences across species to understand evolutionary relationships.
- **Diagnostic Tools:** Utilizing genomic data for the development of diagnostic markers and tests.

EMBL-EBI: A European Resource for Genomic and Proteomic Data

The **European Molecular Biology Laboratory - European Bioinformatics Institute (EMBL-EBI)** is a key player in the bioinformatics community, offering a variety of biological databases and tools. EMBL-EBI provides access to an extensive array of biological data, from **genetic sequences** and **protein structures** to **functional annotations** and **gene expression data.** It is known for its involvement in **international collaborations** and its commitment to making biological data freely accessible to the global scientific community. One of the prominent EMBL-EBI resources is **Ensembl**, which offers a comprehensive genome browser for vertebrate and non-vertebrate species. Ensembl allows researchers to explore not just **genomic sequences** but also **genetic variation, functional annotations**, and **comparative genomics.** EMBL-EBI also hosts **ArrayExpress**, a public repository for high-throughput gene expression data, and **InterPro**, a database that classifies protein families and domains, making it easier to understand protein function.

Features:

- **Comprehensive Databases:** EMBL-EBI hosts a variety of databases covering genomic sequences, functional annotations, protein structures, and more.
- **Integration with Bioinformatics Tools:** EMBL-EBI provides access to powerful bioinformatics tools for sequence analysis, protein modeling, and functional genomics.
- **International Collaboration:** EMBL-EBI plays a key role in global data-sharing initiatives, ensuring that the data is open and accessible.

Applications:

- **Genome Annotation:** Identifying genes, regulatory elements, and functional regions in newly sequenced genomes.
- **Functional Genomics:** Linking gene sequences to their biological functions through expression data and pathway analysis.
- **Protein Function Prediction:** Using protein domain and motif information to predict protein activity.

UniProt: The Comprehensive Protein Database

UniProt (Universal Protein Resource) is one of the most authoritative and comprehensive protein databases available. It provides high-quality, curated information on the **protein sequences, functions, structures, interactions**, and **post-translational modifications** of proteins from a wide range of organisms. UniProt is a collaboration between **EMBL-EBI, Swiss Institute of Bioinformatics (SIB)**, and **Protein Information Resource (PIR)**. UniProt's core product, **UniProtKB (UniProt Knowledgebase)**, consists of two main sections: **Swiss-Prot**, a manually annotated and reviewed section, and **TrEMBL**, an automatically annotated section. UniProtKB offers detailed protein annotations, including information on **protein families, domains, enzyme functions**, and **disease associations**, among other attributes. UniProt is also linked to other databases, such as **PDB** (Protein Data Bank) for structural data and **GO** (Gene Ontology) for functional classification.

Features:

- **Comprehensive Protein Annotations:** Provides information on protein sequences, functions, structures, interactions, and mutations.
- **High-Quality Data:** Swiss-Prot entries are manually curated, ensuring high accuracy and reliability of protein data.
- **Cross-Database Integration:** Links to external resources such as **PDB, GO**, and **Ensembl** for an integrated view of protein data.

Applications:

- **Proteomics Research:** Studying protein expression, structure-function relationships, and post-translational modifications.
- **Disease Association:** Identifying mutations in proteins that contribute to genetic disorders and other diseases.
- **Drug Discovery:** Identifying potential drug targets by understanding protein function and interactions.

Other Key Public Databases

In addition to GenBank, EMBL, and UniProt, there are numerous other specialized and general-purpose public databases that provide critical data for researchers in various biological fields:

- **Ensembl:** A genome database that offers detailed annotations for a wide range of organisms, with powerful tools for visualizing and comparing genomes.
- **dbSNP:** A database focused on **single nucleotide polymorphisms (SNPs)**, cataloging genetic variations that are commonly found in human populations and their associations with diseases.

- **KEGG (Kyoto Encyclopedia of Genes and Genomes):** A comprehensive database for understanding high-level functions of biological systems, including **metabolic pathways**, **signal transduction**, and **disease pathways.**
- **RefSeq:** A curated collection of **genomic sequences**, including **mRNA** and **protein sequences**, providing a reference for gene annotation and functional analysis.
- **The Cancer Genome Atlas (TCGA):** A collection of genomic data specifically focused on various cancers, offering insights into the genetic basis of cancer development and potential therapeutic targets.

Why Public Databases Matter

Public databases like GenBank, EMBL, and UniProt are fundamental to the progression of biological sciences for several reasons:

- **Open Access to Data:** They allow free, unrestricted access to valuable biological data, which accelerates scientific discoveries by enabling researchers worldwide to collaborate and build on each other's work.
- **Standardization and Quality Control:** Many public databases, especially GenBank, EMBL, and UniProt, enforce rigorous standards and protocols for data submission, ensuring that the information stored is accurate and consistent.
- **Fostering Innovation:** Public databases provide a repository for new sequence data, functional annotations, and research findings, supporting innovation in genomics, drug discovery, and clinical research.
- **Large-Scale Collaborative Efforts:** These databases are often central hubs for large-scale, multi-institutional research projects, such as the **Human Genome Project** or **The Cancer Genome Atlas**, allowing global collaboration and data integration across disciplines.

Public biological databases like **GenBank**, **EMBL**, and **UniProt** serve as invaluable resources for the global scientific community. By providing free and open access to genetic, genomic, and proteomic data, these databases empower researchers to explore biological systems in unprecedented detail. The integration of these databases with powerful bioinformatics tools further accelerates discovery, enabling new insights into health, disease, and evolution. As biological research continues to advance, these public databases will remain at the forefront of scientific innovation, driving progress in medicine, biotechnology, and environmental science.

DATABASE QUERYING AND RETRIEVAL

Database querying and retrieval are essential processes in bioinformatics that enable researchers to access, extract, and analyze biological data from large-scale biological databases. These databases often contain vast amounts of genetic, genomic, proteomic, or functional information that researchers need to investigate specific questions, such as identifying gene sequences, finding homologous proteins, or studying genetic variations linked to diseases. Querying involves the use of specialized search tools and query languages, such as SQL (Structured Query Language) or more biologically tailored approaches like BLAST (Basic Local Alignment Search Tool), which allow users to specify search criteria and retrieve relevant data from databases like GenBank, UniProt, or Ensembl. Retrieval can be as simple as a keyword search or as complex as querying for sequences with specific functional annotations, gene expression profiles, or protein interaction data. Effective querying and retrieval empower researchers to perform tasks such as sequence alignment, comparative genomics, disease association studies, and functional annotation, facilitating the discovery of new insights and advancing our understanding of biological systems. With the increasing complexity and size of biological datasets, optimized querying tools and user-friendly interfaces are critical for enabling efficient and accurate data retrieval. In bioinformatics, database querying and retrieval are key steps in accessing and extracting relevant biological information from large, structured biological databases. These processes allow researchers to efficiently find the specific data they need for their studies, whether it's genomic sequences, protein structures, functional annotations, or other biological entities. Given the vast size and complexity of biological datasets, efficient querying and retrieval tools are crucial for narrowing down search results to meaningful and relevant information. In this context, querying refers to the act of submitting a search request to the database based on specified criteria, while retrieval refers to the process of extracting and presenting the results of that query in a useful format.

Types of Biological Data Repositories

Biological databases store a wide range of data types, including:

- **Genomic Data:** Sequences of DNA, RNA, and entire genomes.
- **Proteomic Data:** Information on proteins, their sequences, structures, functions, and interactions.
- **Functional Data:** Annotations linking genes or proteins to biological functions, pathways, and disease associations.
- **Clinical Data:** Genetic variation, disease associations, and epidemiological data.

Different biological databases are optimized for storing specific types of data. For example:

- **GenBank** focuses on storing **nucleotide sequences** from a wide variety of organisms.
- **UniProt** specializes in **protein sequences** and functional annotations.
- **Ensembl** provides genome annotations for vertebrate species and model organisms.

Effective querying and retrieval involve navigating the structure of these databases, which are often organized into complex tables and relationships.

Database Querying

Understanding the Process: Querying a biological database involves specifying search criteria in order to extract data that match particular parameters. Researchers can formulate queries to find:

- **Specific Sequences:** For example, searching for a particular gene in a specific organism or retrieving all sequences with a particular motif.
- **Genetic Variations:** Searching for specific genetic variants, such as **single nucleotide polymorphisms (SNPs)**, that are associated with certain diseases or traits.
- **Functional Annotations:** Finding proteins with specific functional domains or identifying genes that are involved in a particular biological process.

Common Querying Methods

Keyword Search:

- A basic method where users enter keywords related to a biological feature (e.g., gene name, protein function, or disease association) into a search field. The database returns results that match those keywords in its records.
- Example: A search for the term "BRCA1" in **GenBank** will return all sequences that mention or relate to the BRCA1 gene.

Sequence-Based Search:

- More advanced than a simple keyword search, sequence-based queries involve using a DNA, RNA, or protein sequence to search for matches across a database. Tools like **BLAST** allow users to input a sequence and compare it against a vast database of known sequences to find similar or identical sequences.
- Example: A researcher may submit a nucleotide sequence to **GenBank** and use **BLAST** to find other sequences with similar base pair arrangements,

helping them identify homologous genes or study evolutionary relationships.

Advanced Querying with Specific Criteria:

- Many databases allow users to set more complex filters or query parameters. For example, users can specify organism types, sequence lengths, taxonomic classification, or functional annotations to refine their search. In databases like **Ensembl**, users can specify genetic loci, chromosome locations, and even SNPs to retrieve genomic information.
- Example: A query to **Ensembl** for genes located on human chromosome 1 involved in immune responses can return a set of gene sequences that match these criteria.

Predefined Query Templates:

- Some databases, such as **UniProt**, provide predefined templates or advanced search options that let users filter their search by specific protein attributes, such as molecular weight, subcellular location, biological processes, or disease associations.
- Example: A query might search for all human proteins that are involved in "DNA repair" and also contain a specific protein domain.

Database Retrieval: Extracting and Interpreting Results

Once a query is submitted, the **retrieval** process refers to the extraction and presentation of the relevant data from the database. Retrieved data is typically displayed in a user-friendly format, such as tables, lists, or interactive visualizations, that makes it easy to interpret and analyze. The quality and clarity of the retrieval interface are essential for effective research, as it helps researchers quickly assess the relevance of the retrieved data.

Key aspects of the retrieval process include:

- **Data Presentation:** Results are typically presented with links to related information, such as gene annotations, protein domains, related sequences, or references to scientific literature.
- **Sequence Alignment and Visualization:** Many biological databases integrate visualization tools (e.g., **Genome Browsers, Protein Structure Viewers**) that allow users to view the retrieved sequences or structures in an interactive format. For instance, **Ensembl** offers an interactive genome browser where researchers can view retrieved genomic data in the context of the entire chromosome.

- **Data Export:** After querying and retrieving data, many databases allow users to export the results in various formats (e.g., FASTA, CSV, Excel, or XML) for further analysis or integration with other software tools.

Common Tools for Querying and Retrieval in Bioinformatics

Several bioinformatics tools and interfaces support the querying and retrieval of data from biological databases. Some of the most popular tools include:

BLAST (Basic Local Alignment Search Tool):

- BLAST is one of the most widely used sequence search tools for comparing a query sequence against a database of known sequences. It identifies regions of similarity and can provide insights into gene homology, functional annotations, and evolutionary relationships.
- Example: Researchers might use **BLASTn** to search a nucleotide query sequence against **GenBank** or **BLASTp** to search for similar protein sequences.

FASTA Format:

- A common format for storing and retrieving biological sequence data. Many databases allow users to download retrieved sequences in FASTA format for further analysis using bioinformatics tools.

SQL and NoSQL Queries:

- Many biological databases, especially large-scale genome or multi-dimensional datasets, are based on relational databases or NoSQL databases, where users can perform complex queries using **SQL (Structured Query Language)** or other query languages. These queries can be used to filter data based on multiple criteria such as gene expression levels, sequence length, or specific variants.

Web Interfaces and API Access:

- Most modern biological databases offer **web interfaces** that allow users to perform queries without requiring advanced programming knowledge. In addition, **Application Programming Interfaces (APIs)** are available for programmatic access to databases, allowing researchers to automate data retrieval and integrate biological data into their workflows.

Challenges in Database Querying and Retrieval

While querying and retrieval are critical steps, they also present several challenges:

- **Data Volume and Complexity:** As biological data continues to grow exponentially, efficient querying becomes increasingly complex. Users need advanced tools to manage, filter, and query large datasets effectively.

- **Data Quality and Consistency:** Databases must maintain high-quality, accurate, and consistent data. Inconsistent or poorly annotated data can lead to incorrect conclusions or missed findings.
- **Integration Across Databases:** Biological data often comes from multiple databases, each storing different aspects of the same biological entity (e.g., a gene sequence in GenBank, functional annotations in UniProt, and pathway information in KEGG). Combining data from various sources to get a comprehensive view of a biological question requires careful integration and cross-database querying.

Applications of Database Querying and Retrieval

The ability to effectively query and retrieve data from biological databases underpins numerous bioinformatics applications:

- **Genomic Research:** Identifying novel genes, regulatory elements, and mutations by comparing newly sequenced genomes with existing genomic data.
- **Comparative Genomics:** Analyzing evolutionary relationships by comparing gene sequences across different species.
- **Proteomics:** Investigating protein functions, structures, and interactions to understand cellular processes and disease mechanisms.
- **Disease Research:** Using genetic and variant data to identify disease-associated genes or mutations and understand their role in disease pathogenesis.

Database querying and retrieval are fundamental processes in bioinformatics that empower researchers to extract meaningful insights from large biological datasets. By enabling users to search for specific sequences, functional annotations, and genetic variants, querying and retrieval tools help accelerate discovery in fields ranging from genomics and proteomics to disease research and drug development. With the continuous growth of biological data, developing and refining efficient, user-friendly querying tools remains a critical challenge for the bioinformatics community. These tools will continue to evolve, enabling more sophisticated, large-scale analyses of biological data and contributing to the advancement of personalized medicine, biotechnology, and molecular biology.

SPECIALIZED DATABASES FOR NUCLEIC ACIDS, PROTEINS AND PATHWAYS

Specialized databases are essential in bioinformatics and molecular biology for storing and providing access to highly specific types of biological data. These databases are focused on particular aspects of molecular biology, such as nucleic acids (DNA and RNA), proteins, and biological pathways. Each specialized database typically curates data with a deep focus on one type of biomolecule, offering researchers the ability to query, retrieve, and analyze detailed information related to the structure, function, and interactions of nucleic acids and proteins, or the regulatory networks in which they are involved.

Specialized Databases for Nucleic Acids

Nucleic acid databases primarily store and provide access to DNA and RNA sequences. These databases are critical for genomics, gene expression studies, and RNA sequencing, among other applications. Many of these resources are dedicated to specific types of nucleic acid information, such as sequence data, functional annotations, and variation data.

Examples of Nucleic Acid Databases:

GenBank (NCBI)

- **Focus:** GenBank is one of the largest public repositories for nucleotide sequences and provides access to a wide variety of sequences from different organisms, including genomic DNA, mRNA, and non-coding RNA sequences.
- **Data Content:** Includes both raw sequence data and annotated sequence records.
- **Applications:** GenBank is essential for tasks like sequence alignment, gene discovery, functional annotation, and comparative genomics.
- **Linkage to Other Databases**: GenBank is linked to other databases such as RefSeq (for curated reference sequences) and dbSNP (for genetic variations).

Ensembl

- **Focus:** Ensembl provides genome-wide data for multiple species, with detailed information on gene annotations, variation, transcripts, and regulatory elements.
- **Data Content:** Genomic sequences, including mRNA, DNA, and coding/non-coding regions, are available for a large number of organisms.
- **Applications:** Ensembl is particularly useful for genome browser queries, comparative genomics, and studying gene evolution.
- **Tools:** Includes powerful visualization tools for visualizing genomic data, SNPs, and functional elements in the context of entire genomes.

The RNAcentral Database

- **Focus:** RNAcentral provides comprehensive data on non-coding RNA sequences from various organisms.

- **Data Content:** Includes information on miRNA, siRNA, rRNA, and other small RNAs.
- **Applications:** Researchers use RNAcentral for RNA annotation, studying RNA secondary structure, and identifying novel non-coding RNA genes.

dbSNP (NCBI)

- **Focus:** dbSNP is a database specifically for storing information on single nucleotide polymorphisms (SNPs), which are the most common type of genetic variation in the human genome.
- **Data Content:** SNPs, indels, and other types of genetic variation, along with their functional effects and associated diseases.
- **Applications:** dbSNP is used in genetic association studies, genome-wide association studies (GWAS), and personalized medicine.

Specialized Databases for Proteins

Protein databases are designed to store and organize data related to the sequences, structures, functions, and interactions of proteins. Given the central role of proteins in biological systems, these databases are indispensable for researchers in proteomics, functional genomics, and drug discovery.

Examples of Protein Databases:

UniProt (Universal Protein Resource)

- **Focus:** UniProt is one of the most comprehensive protein databases available, providing detailed and curated information on protein sequences, structures, functions, and diseases.
- **Data Content:** Manually curated protein entries (Swiss-Prot), automatically annotated entries (TrEMBL), and information about protein functions, interactions, pathways, and post-translational modifications.
- **Applications:** UniProt is widely used for protein function prediction, functional annotation, comparative proteomics, and drug target discovery.
- **Linkage to Other Databases:** UniProt is integrated with PDB (Protein Data Bank), GO (Gene Ontology), and InterPro (protein domain and family database).

Protein Data Bank (PDB)

- **Focus:** The PDB is a specialized database that provides 3D structural data of proteins, nucleic acids, and complex biomolecular structures.
- **Data Content:** High-resolution 3D structures of proteins and other biomolecules determined through X-ray crystallography, NMR spectroscopy, or cryogenic electron microscopy (cryo-EM).

- **Applications:** Used for structural bioinformatics, drug design, and molecular docking studies.
- **Tools:** The PDB offers visualization tools like Jmol and PyMOL to explore and analyze protein structures in 3D.

InterPro

- **Focus:** InterPro provides an integrated resource for the classification and functional annotation of proteins.
- **Data Content:** Information on protein families, domains, motifs, and functional sites, as well as links to other databases such as Pfam, SMART, and ProDom.
- **Applications:** InterPro is used to predict the functions of uncharacterized proteins, identify protein families, and perform functional genomics analyses.

PANTHER (Protein ANalysis THrough Evolutionary Relationships)

- **Focus:** PANTHER is a database that classifies proteins based on their evolutionary relationships and functional annotations.
- **Data Content:** It includes protein families, pathways, and ontologies (biological processes, molecular functions, and cellular components).
- **Applications:** PANTHER is frequently used for gene set enrichment analysis and functional analysis of high-throughput data, such as RNA-Seq.

Specialized Databases for Biological Pathways

Biological pathway databases focus on the molecular networks that govern cellular processes and biological functions, such as metabolism, signal transduction, and gene regulation. These databases allow researchers to study how various molecules (genes, proteins, metabolites) interact and influence each other within complex biological systems.

Examples of Pathway Databases:

KEGG (Kyoto Encyclopedia of Genes and Genomes)

- **Focus:** KEGG is one of the most comprehensive and widely used pathway databases. It provides detailed maps of metabolic pathways, signal transduction pathways, and disease pathways, along with associated genomic data.
- **Data Content:** Pathways, enzyme reactions, genes, small molecules, and diseases. KEGG also offers tools to visualize and analyze metabolic networks, pathway enrichment, and gene function.

- **Applications:** KEGG is used in metabolomics, systems biology, and drug discovery, helping to identify potential therapeutic targets or biomarkers based on pathway analysis.
- **Tools:** KEGG Mapper is a tool that allows users to map their gene or protein data onto pathway maps.

Reactome

- **Focus:** Reactome is a free, open-source database of biological pathways that covers a wide range of processes such as cell cycle regulation, signal transduction, metabolism, and gene expression.
- **Data Content:** Curated pathway data, including molecular interactions, biochemical reactions, and regulatory processes.
- **Applications:** Reactome is widely used for pathway enrichment analysis, gene set analysis, and visualizing interactions between molecules.
- **Tools:** Reactome provides interactive pathway diagrams and integrates with other bioinformatics tools like Cytoscape for network visualization.

WikiPathways

- **Focus:** WikiPathways is a community-driven platform that offers a vast collection of biological pathway diagrams, contributed by researchers from around the world.
- **Data Content:** Pathway diagrams, with annotations linking genes, proteins, and other molecules to biological functions and processes.
- **Applications:** WikiPathways is used for pathway curation, disease modeling, and network analysis in systems biology.
- **Tools:** WikiPathways offers integration with other databases like Ensembl, Reactome, and KEGG, allowing for pathway enrichment and functional analysis.

BioCarta

- **Focus:** BioCarta provides pathways focused on key signaling and metabolic networks that are relevant to human health and disease.
- **Data Content:** A collection of pathway diagrams primarily related to signal transduction, immune response, and cancer.
- **Applications:** Used for target identification, biomarker discovery, and drug development, particularly in the context of cancer, diabetes, and cardiovascular diseases.

6
Chapter
SEQUENCE ANALYSIS

INTRODUCTION

Sequence analysis refers to the process of examining and interpreting biological sequences, such as DNA, RNA, or protein sequences, to derive meaningful insights about their structure, function, and evolutionary history. This approach is a cornerstone of modern bioinformatics and computational biology, enabling researchers to study the genetic and molecular bases of life. The primary goal of sequence analysis is to understand the relationships between sequences, identify conserved regions that may be important for biological function, and uncover evolutionary patterns. Techniques such as sequence alignment, motif discovery, and phylogenetic tree construction are commonly used in sequence analysis to compare sequences across different species, or to predict the structure and function of newly discovered sequences. With the advent of high-throughput sequencing technologies, the volume of sequence data has exploded, making computational methods an essential tool for managing, analyzing, and interpreting this wealth of information. Sequence analysis not only enhances our understanding of genetics but also facilitates advancements in medicine, agriculture, and biotechnology by identifying disease-causing mutations, discovering new drug targets, and improving crop traits. In this context, sequence analysis is not merely a technical skill but a powerful lens through which the complexities of biological systems can be explored and understood.

SEQUENCE ALIGNMENT (PAIRWISE AND MULTIPLE SEQUENCE ALIGNMENT)

Sequence alignment is a fundamental technique in bioinformatics that involves arranging two or more biological sequences (such as DNA, RNA, or protein sequences) to identify regions of similarity. These similarities can be indicative of functional, structural, or evolutionary relationships between the sequences. Sequence alignment is divided into two main types: pairwise sequence alignment and multiple sequence alignment. Each serves a different purpose and involves distinct methodologies and computational approaches.

Pairwise Sequence Alignment

Pairwise sequence alignment compares two sequences (e.g., two DNA, RNA, or protein sequences) to identify regions where they are similar or identical. This can provide insights into evolutionary relationships, functional similarities, and shared characteristics.

Concepts:

- **Matches:** When two sequences have identical nucleotides or amino acids at the same position.
- **Mismatches:** When the nucleotides or amino acids at a given position are different between the two sequences.
- **Gaps:** Insertions or deletions (indels) in one sequence relative to the other.

Pairwise alignment is often divided into global alignment and local alignment:

- **Global Alignment:** Involves aligning the entire length of both sequences from start to end. This is useful when the sequences are of similar length and have a high degree of overall similarity. The most common algorithm used for global alignment is the Needleman-Wunsch algorithm.
- **Local Alignment:** Focuses on aligning the most similar subregions (or local segments) within the two sequences. This is ideal for situations where the sequences may have regions of strong similarity, but differ significantly overall. The Smith-Waterman algorithm is commonly used for local alignment.

Example of Pairwise Sequence Alignment:

Consider two DNA sequences:

Sequence 1: AGCTGAC Sequence 2: AGTTGAC

The goal is to align these sequences by introducing gaps and matching nucleotides where possible. The pairwise alignment might look like this:

AGCTGAC AG-TGAC

In this example, a gap is introduced in sequence 2 to optimize the alignment and maximize the matching bases.

Algorithms:

- **Needleman-Wunsch Algorithm:** A dynamic programming algorithm used for global pairwise sequence alignment. It constructs a matrix of scores based on matches, mismatches, and gaps, and then traces back through the matrix to find the optimal alignment.

- **Smith-Waterman Algorithm:** A dynamic programming algorithm for local pairwise sequence alignment. It similarly constructs a scoring matrix but allows for gaps and mismatches to be excluded in the final alignment, focusing on the best matching local subsequence.

Multiple Sequence Alignment (MSA)

Multiple sequence alignment (MSA) extends the concept of pairwise alignment to align more than two sequences simultaneously. MSA is crucial for studying evolutionary relationships, identifying conserved motifs, and predicting the function of proteins or genes across different species.

Concepts:

- **Conserved Regions:** In MSA, conserved regions across multiple sequences are often of particular interest, as these may indicate functionally important sites.
- **Gaps:** Gaps are inserted in sequences to maximize alignment, particularly when sequences diverge over evolutionary time.

MSA can be more challenging than pairwise alignment because it involves aligning multiple sequences simultaneously, leading to more complex patterns of matches, mismatches, and gaps. Moreover, the computational cost increases with the number of sequences and their length.

Algorithms for MSA:

- **ClustalW:** One of the most widely used algorithms for multiple sequence alignment. ClustalW uses a progressive alignment approach where sequences are aligned pairwise in a stepwise manner, and the best alignments are then combined to build the final multiple sequence alignment.
- **MAFFT:** An alternative to ClustalW, MAFFT is a more computationally efficient tool, particularly for large datasets. It uses a fast Fourier transform (FFT) to accelerate the alignment process, making it scalable for large numbers of sequences.
- **T-Coffee:** T-Coffee improves upon traditional methods by combining results from several alignment methods and optimizing the final MSA. This approach is often used for higher accuracy in the presence of complex or divergent sequences.
- **MUSCLE:** MUSCLE (Multiple Sequence Comparison by Log-Expectation) is a program for fast and accurate multiple sequence alignment, often outperforming ClustalW in terms of speed and accuracy.

Example of Multiple Sequence Alignment:

Consider three protein sequences:

Sequence 1: AGCTGAC Sequence 2: AG-TGAC Sequence 3: AGT-TGC

A multiple sequence alignment might look like this:

AGCTGAC AG-TGAC AGT-TGC

Here, gaps are introduced in sequence 2 and sequence 3 to optimize the alignment, and conserved regions (e.g., "AGT") are identified across all sequences.

Applications of MSA:

- **Phylogenetic Tree Construction:** By aligning sequences from different species or organisms, one can infer the evolutionary relationships between them.
- **Motif Identification:** Conserved motifs or domains within proteins can be identified by aligning homologous sequences from different species or organisms.
- **Functional Annotation:** By aligning a newly sequenced gene or protein with known sequences, researchers can predict the function of the newly sequenced molecule based on the similarity of its sequence to previously characterized genes or proteins.

Challenges in MSA:

- **Handling of Divergent Sequences:** When sequences are highly divergent, it becomes more difficult to identify conserved regions, and alignment algorithms may struggle to produce meaningful results.
- **Computational Complexity:** Aligning many sequences with large lengths can be computationally intensive, requiring efficient algorithms and considerable processing power.

Differences Between Pairwise and Multiple Sequence Alignment:

- Number of Sequences: Pairwise alignment compares two sequences, whereas multiple sequence alignment compares three or more sequences simultaneously.
- Complexity: Pairwise alignment is generally less computationally complex than multiple sequence alignment. MSA requires more sophisticated algorithms to handle the complexity of aligning multiple sequences and to ensure the most accurate alignment.
- Purpose: Pairwise alignment is often used for comparing individual sequences (e.g., finding mutations or evolutionary relationships between

two species), while MSA is used to study broader evolutionary trends, functional regions, or protein domains across many sequences.

Sequence alignment, both pairwise and multiple, is an essential tool in bioinformatics, enabling researchers to uncover valuable insights into molecular evolution, function, and structure. Pairwise sequence alignment is useful for direct comparisons of two sequences, while multiple sequence alignment provides a more comprehensive view of relationships across multiple sequences. The choice of method depends on the research question and the number and nature of the sequences involved. As sequence data continues to grow exponentially, these alignment techniques remain central to understanding the complexities of the genetic and protein worlds.

SEQUENCE SIMILARITY AND IDENTITY

In the context of sequence analysis, the terms **sequence similarity** and **sequence identity** are often used to compare biological sequences (such as DNA, RNA, or protein sequences). Both concepts refer to the degree of match between two sequences, but they differ in how they are calculated and what they represent biologically.

Sequence Identity

Sequence identity refers to the percentage of exact matches between two aligned sequences. It is a straightforward measure of similarity, focusing purely on the positions where the same nucleotides or amino acids appear in both sequences.

- **DNA/RNA Identity**: For nucleotide sequences, identity is determined by counting the number of positions where the same base (A, T, G, or C in DNA) or nucleotide (A, U, G, or C in RNA) appears in both sequences. A perfect match at a position is counted as a "hit."
- **Protein Identity**: For protein sequences, identity is based on matching amino acids at corresponding positions. Proteins are made up of 20 different amino acids, and identity counts how often the same amino acid appears at a given position in both sequences.

Formula for Sequence Identity: Sequence Identity (%)=Number of Identical MatchesTotal Number of Positions in the Alignment×100Sequence Identity (%)=Total Number of Positions in the AlignmentNumber of Identical Matches×100

Example of Sequence Identity:

Consider two DNA sequences:

Sequence 1: ATCGGTA Sequence 2: ATCGCTA

Aligning these sequences results in the following:

ATCGGTA ATCGCTA

In this case:

- Matches: A (at position 1), T (at position 2), C (at position 3), G (at position 4), and A (at position 7).
- Mismatch: The fifth and sixth positions have a G in Sequence 1 and a C in Sequence 2.

Here, there are 5 exact matches and 2 mismatches out of 7 total positions.

Sequence Identity

Identity=57×100≈71.43%Identity=75×100≈71.43%

In this case, the identity between the two sequences is **71.43%.**

Sequence Similarity

Sequence similarity, on the other hand, takes into account both exact matches (identity) and similar residues. When calculating similarity, the concept is extended to include substitutions where the substituted residues are chemically or functionally similar. For example:

- In DNA, purine (A, G) and pyrimidine (C, T) bases are often considered functionally similar.
- In proteins, certain amino acids with similar side chains (e.g., lysine and arginine, or alanine and valine) might be counted as "similar" because they have similar chemical properties.

Unlike identity, similarity reflects how alike two sequences are in terms of their biochemical or structural properties, even if they do not share an exact match at every position.

Sequence Similarity Scoring

To quantify sequence similarity, scoring matrices like PAM (Point Accepted Mutation) or BLOSUM (BLOcks SUbstitution Matrix) are often used, particularly for protein sequences. These matrices assign scores to each possible pair of amino acids, reflecting their likelihood of being substituted for each other during evolution.

For example:

- **PAM matrices** provide scores based on the evolutionary distance between two amino acids, with positive scores for substitutions that are more likely and negative scores for unlikely substitutions.
- **BLOSUM matrices** are based on observed substitutions in conserved regions of protein families, and the values are derived from sequence alignments of protein blocks.

Formula for Sequence Similarity

The similarity score can be calculated based on a substitution matrix, where identical residues score the highest (positive), and mismatched residues or less similar substitutions score lower (often negative). The formula for similarity varies depending on the scoring system and the type of sequences (DNA, RNA, or protein) involved.

Example of Sequence Similarity:

Consider two protein sequences:

Sequence 1: MKTAYIAKQRQISFVL Sequence 2: MKTAYVAKQRQISFML

When aligned, these sequences show:

- Matches (exact identity): M (1), K (2), T (3), A (4), Y (5), I (6), A (7), K (8), Q (9), R (10), Q (11), I (12), S (13), F (14), V (15)
- Similar substitutions: A (at position 6 in Sequence 1 and V at position 6 in Sequence 2; A and V are somewhat similar because both are hydrophobic amino acids).
- Mismatches: L (at position 16 in Sequence 1) and M (at position 16 in Sequence 2).

In a scoring matrix for proteins (e.g., BLOSUM62), similar residues like A and V may receive a positive score, while mismatches will be penalized with a negative score.

Differences Between Sequence Identity and Similarity

Characteristic	Sequence Identity	Sequence Similarity
Definition	Percentage of exact matches between two sequences.	Degree of similarity, considering both exact matches and functionally similar substitutions.
Calculation	Number of exact matching positions divided by the total number of positions.	A weighted sum that includes exact matches and similar substitutions based on scoring matrices (e.g., BLOSUM, PAM).
Use Case	Used to measure the exact degree of correspondence between two sequences.	Used to assess the functional or evolutionary relationship between two sequences.
Nature of Comparison	Only considers exact matches (e.g., A=A or G=G).	Considers both exact matches and similar residues (e.g., A=V, K=R).
Scope	Often used for high-identity comparisons where precision is important (e.g., detecting mutations).	Useful for evolutionary studies or when comparing sequences with some degree of divergence.

Applications and Importance

Sequence Identity:

- **Mutation Detection:** Sequence identity is particularly useful in detecting mutations or assessing the genetic difference between two sequences. High identity indicates closely related sequences, while low identity may suggest significant genetic divergence.
- **Exact Matching:** When comparing sequences that are expected to be highly conserved (e.g., within a species or closely related species), identity is a useful metric for determining how closely the sequences match.

Sequence Similarity:

- **Evolutionary Studies:** Similarity is used to infer evolutionary relationships. Sequences that are functionally or structurally similar (even if they are not identical) likely share a common ancestor.
- **Protein Structure and Function:** When comparing protein sequences, similarity can help identify conserved functional regions, even if exact amino acid residues are not conserved across species.
- **Homology Modeling:** In structural biology, sequence similarity is key for predicting the 3D structure of proteins based on homologous sequences (proteins with similar sequences).

Both **sequence identity** and **sequence similarity** are essential concepts in bioinformatics, each serving different purposes. Sequence identity provides a direct, precise measure of how closely two sequences match, while sequence similarity offers a broader view by incorporating functionally or structurally related residues, even when exact matches are not present. Understanding these two concepts and how to compute them is crucial for a wide range of applications in genomics, proteomics, evolutionary biology, and molecular biology.

SUBSTITUTION MATRICES (E.G., PAM, BLOSUM)

Substitution matrices are fundamental tools in sequence alignment, particularly when comparing protein sequences. They provide a way to score substitutions between amino acids, helping to quantify the degree of similarity or dissimilarity between sequences during the alignment process. These matrices are essential for understanding how likely one amino acid is to be replaced by another during evolutionary processes. Two of the most widely used substitution matrices are PAM (Point Accepted Mutation) and BLOSUM (BLOcks Substitution Matrix). These matrices are based on different methodologies and assumptions, and each has its specific applications and advantages.

PAM (Point Accepted Mutation) Matrix

The PAM matrix was developed by Margaret Dayhoff and colleagues in the 1970s. It is based on the concept of point accepted mutations, which are the mutations that have become fixed in a population over time. The PAM matrix quantifies how likely one amino acid is to mutate into another over a certain evolutionary distance, typically measured in PAM units.

Concepts:

- **PAM Unit:** A PAM unit represents a mutation that occurs once per 100 amino acids. A **PAM1 matrix** refers to mutations that occur over a distance of 1 PAM unit, meaning that for every 100 amino acids, one mutation is expected to occur.
- **Evolutionary Distance:** PAM matrices are based on evolutionary relationships between proteins. The further apart two species are in terms of evolutionary time, the larger the PAM number required to measure the divergence between their sequences.
- **Mutation Rates:** PAM matrices are derived from observed substitutions between amino acids in homologous protein sequences. The substitution probabilities reflect the likelihood of one amino acid being replaced by another due to evolutionary pressures, including structural constraints, chemical properties, and functional requirements.

PAM Matrix Construction:

PAM matrices are built by aligning a large number of homologous protein sequences and calculating the frequencies with which certain amino acids are substituted for each other. This is done for sequences of known evolutionary distance (e.g., 1 PAM, 100 PAM, etc.). The most commonly used PAM matrices are PAM1, PAM30, and PAM250.

- **PAM1:** Measures very closely related proteins (about 1% divergence).
- **PAM30:** Used for proteins with about 30% divergence.
- **PAM250:** Used for sequences that are very divergent (greater than 50% divergence), commonly used in more distantly related species comparisons.

Example of PAM1 Substitution Matrix:

Here's a simplified example of a PAM1 matrix for a few amino acids (showing only a small section):

	A	C	G	T
A	5	-1	-1	-1
C	-1	9	-3	-3
G	-1	-3	7	-2
T	-1	-3	-2	5

- The diagonal values represent the **match score**, indicating the likelihood of an amino acid being substituted by itself (e.g., A → A, G → G).
- The off-diagonal values represent the **substitution scores** for different amino acids (e.g., A → C, A → G, etc.).

The positive values indicate that the substitution is favored, while negative values indicate less favorable substitutions.

PAM Matrices for Different Evolutionary Distances:

- **PAM1:** Small mutations, mainly used for closely related sequences.
- **PAM250:** Large mutations, used for distantly related sequences.
- PAM matrices are **generally more suited for closely related sequences** because they assume that the mutation rate is small and proteins are evolving gradually.

BLOSUM (BLOcks Substitution Matrix): The **BLOSUM** matrix, developed by **Henikoff and Henikoff** in 1992, stands for **BLOcks Substitution Matrix.** Unlike PAM, which is based on evolutionary distances and calculates substitution probabilities based on the assumption of a gradual accumulation of mutations, BLOSUM is based on **block-based alignments** of sequences.

Concepts:

- **BLOSUM Score**: BLOSUM matrices are based on the frequency of observed substitutions in conserved regions (blocks) of aligned sequences, often focusing on highly conserved functional regions across multiple species.
- **BLOSUM Number**: The number associated with a BLOSUM matrix (e.g., BLOSUM62, BLOSUM80) refers to the threshold for sequence identity used to create the matrix. For example, BLOSUM62 is derived from sequences that have at least 62% sequence identity, meaning it is designed for use with sequences that are moderately similar.
- **Block Alignment**: BLOSUM matrices are built by aligning blocks of conserved protein sequences, rather than by directly comparing full-length sequences. These blocks are highly conserved regions within protein families.

BLOSUM Matrix Construction:

BLOSUM matrices are derived by examining aligned sequences of proteins from multiple species. The alignment is done by considering only the most conserved regions (blocks) and determining the frequency of specific amino acid substitutions within those blocks. The substitution scores in a BLOSUM matrix reflect the observed likelihood of substitution in conserved protein regions.The BLOSUM matrices are designed to be more effective for sequences with varying degrees of similarity, especially in distantly related proteins. The most commonly used BLOSUM matrix is BLOSUM62, which strikes a balance between being effective for both closely and distantly related sequences.

Example of BLOSUM62 Substitution Matrix:

Here's a simplified example of the BLOSUM62 matrix (again, a small section):

	A	C	G	T
A	4	-1	-2	-1
C	-1	9	-3	-2
G	-2	-3	6	-2
T	-1	-2	-2	5

- The values on the diagonal represent the substitution of an amino acid with itself, just like the PAM matrix.
- Off-diagonal values represent the scores for substitutions between different amino acids.

Differences from PAM:

- **BLOSUM matrices are based on real sequence data** from conserved blocks, while PAM matrices are based on evolutionary distances and mutation rates.
- **BLOSUM matrices tend to work better for distantly related sequences** because they consider a wider variety of substitutions that occur in conserved protein regions.
- BLOSUM matrices do not rely on a fixed evolutionary model and instead use real sequence data, making them more appropriate for general use.

BLOSUM Matrices for Different Sequence Identities:

- **BLOSUM62:** Typically used for sequences with 62% sequence identity.
- **BLOSUM80:** Used for sequences with higher sequence identity, typically for closely related proteins.
- **BLOSUM45:** Used for more distantly related sequences (with lower sequence identity).

Differences Between PAM and BLOSUM Matrices

Feature	PAM	BLOSUM
Basis	Derived from evolutionary distance and mutation rates.	Derived from conserved blocks of aligned sequences.
Calculation	Considers gradual accumulation of mutations over time.	Focuses on actual observed substitutions in conserved regions.
Distance Units	Measured in PAM units (e.g., PAM1, PAM250).	Measured by sequence identity threshold (e.g., BLOSUM62, BLOSUM80).
Use Case	Best for closely related sequences (PAM1, PAM30).	Best for sequences with varying degrees of similarity (e.g., BLOSUM62 for moderate similarity).
Matrix Size	Typically smaller matrices for closely related sequences.	Larger matrices are available for sequences with different evolutionary distances.
Evolutionary Context	Better for understanding small evolutionary changes.	Better for handling large evolutionary differences and conserved regions.

Applications of Substitution Matrices

- **Sequence Alignment**: Both PAM and BLOSUM matrices are used to score alignments in tools like BLAST, FASTA, and others. They help determine the best alignment by assigning higher scores to more similar (or identical) amino acids and penalizing less likely substitutions.
- **Phylogenetic Analysis**: Substitution matrices are important for constructing evolutionary trees and understanding the relationships between proteins and species.
- **Protein Structure and Function Prediction**: By aligning sequences with known structures, substitution matrices help in predicting the structure and function of novel proteins.
- **Homology Search**: These matrices are used to identify homologous proteins by comparing newly sequenced proteins to databases of known sequences.

Substitution matrices like PAM and BLOSUM play a crucial role in the field of bioinformatics, particularly in sequence alignment and evolutionary studies. While PAM matrices are better suited for closely related sequences and evolutionary studies involving small mutations, BLOSUM matrices are more useful for aligning sequences with varying degrees of similarity, especially in more distantly related sequences. Understanding the differences and appropriate applications of these matrices is essential for making accurate biological inferences and for performing efficient sequence comparisons.

7 Chapter

GENOME ASSEMBLY AND ANNOTATION

INTRODUCTION

Genome assembly and annotation are fundamental steps in understanding the genetic makeup of organisms, enabling insights into their biology, evolution, and potential applications in medicine, agriculture, and biotechnology. Genome assembly refers to the process of reconstructing the complete DNA sequence of an organism's genome from fragmented pieces of raw sequencing data. This is achieved through computational methods that piece together short reads, generated by high-throughput sequencing technologies, into longer contiguous sequences called contigs and scaffolds. The accuracy and completeness of the assembly directly impact downstream analyses such as gene prediction, functional annotation, and comparative genomics. Once the genome has been assembled, the next critical step is genome annotation, which involves identifying and marking key genomic elements, such as genes, regulatory regions, and non-coding sequences. Annotation provides context to the genome, helping to assign biological meaning to the sequences by predicting the function of genes, their interactions, and their roles in cellular processes. Annotation also enables the comparison of genomes across species, aiding in the identification of conserved and unique features that may be linked to specific traits or evolutionary adaptations. Together, genome assembly and annotation form the backbone of modern genomics, unlocking vast potential for scientific discovery and applied research.

WHOLE GENOME SEQUENCING AND ASSEMBLY

Whole genome sequencing (WGS) is a comprehensive method used to determine the complete DNA sequence of an organism's genome. This process involves reading the entire genetic material in an organism's chromosomes, providing detailed information about both coding and non-coding regions of the genome. WGS is a powerful tool in genomics and has applications in medicine, agriculture, evolutionary biology, and personalized genomics. Genomic Assembly is the process that follows sequencing and involves reconstructing the complete genome from the raw sequencing data. Because modern sequencing technologies produce many short fragments of DNA, assembly algorithms are needed to piece these fragments back together in the correct order and orientation.

Whole Genome Sequencing (WGS)

Steps in WGS:

- **Sample Collection:** The first step is obtaining a sample that contains the DNA. This can be blood, tissue, saliva, or other biological samples, depending on the organism.
- **DNA Extraction:** DNA is isolated from the sample, and its integrity and quality are assessed to ensure it's suitable for sequencing.
- **Library Preparation:** The DNA is fragmented into smaller pieces (typically 150–600 base pairs for next-generation sequencing methods). Adapters are added to the ends of these fragments to facilitate sequencing.
- **Sequencing:** The DNA fragments are sequenced using one of several high-throughput sequencing technologies (e.g., Illumina, PacBio, Oxford Nanopore). These methods generate millions of short DNA reads.
- **Data Generation:** Sequencers generate raw data, which are typically in the form of short reads (sequences of nucleotides), often ranging from 75–300 base pairs.

Sequencing Technologies:

- **Illumina Sequencing:** This is the most widely used method, which employs sequencing by synthesis. It's high-throughput, produces short reads (typically 150–300 bp), and has high accuracy.
- **PacBio (SMRT) Sequencing:** This method generates long reads (10,000–100,000 bp), making it suitable for assembling large genomes and resolving complex regions such as repeats and structural variants.
- **Oxford Nanopore:** This technology offers real-time sequencing with long reads (10,000–2 million bp), useful for sequencing entire genomes quickly and on-site.
- **10x Genomics:** This technology provides a unique approach for sequencing complex regions and structural variants through barcoding and linked-reads technology.

Genomic Assembly

After sequencing, the fragmented DNA sequences need to be reassembled to reconstruct the original genome. This process can be challenging because the reads are short and there can be errors, gaps, or repetitive sequences.

Assembly Strategies:

- **De Novo Assembly**: This approach involves assembling the genome from scratch without using any reference genome. It relies solely on the sequence data generated by the sequencer.
- **Reference-Based Assembly**: This method uses a known reference genome as a guide. The short reads are aligned to the reference genome, and differences are noted (such as mutations, insertions, deletions, etc.).
- **Hybrid Assembly**: This combines short reads (e.g., from Illumina) with long reads (e.g., from PacBio or Oxford Nanopore) to improve the accuracy and completeness of the assembly. The long reads help resolve complex regions, while short reads offer high accuracy.

Assembly Algorithms:

- **De Bruijn Graph-based Algorithms**: These algorithms (e.g., Velvet, SOAPdenovo) break DNA sequences into smaller subsequences (k-mers) and use these to build a graph that represents the genome.
- **Overlap-Layout-Consensus (OLC) Algorithms**: These algorithms (e.g., Canu, Flye) work by finding overlaps between long reads and aligning them into a consensus sequence.
- **String Graphs**: These algorithms (e.g., SPAdes) represent sequences as strings of characters and look for common substrings to form larger contigs.

Contigs, Scaffolds, and Gaps:

- **Contigs**: Contigs are contiguous sequences of DNA that have been successfully assembled from the short reads. They represent parts of the genome.
- **Scaffolds**: Scaffolds are larger structures that are formed by connecting contigs with the help of paired-end reads or long-range sequencing information.
- **Gaps**: Gaps may remain in the assembly, particularly in regions of high complexity, such as repeats or GC-rich areas. These gaps are often addressed by using long-read technologies or filling strategies.

Post-Assembly Analysis

After assembling the genome, the resulting assembly is typically subjected to a series of quality checks:

- **Quality Assessment**: Tools like **QUAST** are used to assess the completeness and accuracy of the assembly. This includes checking metrics like N50 (a measure of contig size distribution) and the total genome size.

- **Annotation:** Once the genome is assembled, it needs to be annotated. This means identifying genes, regulatory elements, and other functional features in the genome. This is typically done using software like **AUGUSTUS**, **GenBank**, or **GeneMark**.
- **Variant Calling:** If comparing multiple genomes or looking for differences between a reference and the sequenced genome, variant calling is performed to identify SNPs (single nucleotide polymorphisms), insertions, deletions, and structural variants.

Applications of Whole Genome Sequencing

Medical Genomics:

- Identifying genetic diseases and mutations.
- Personalized medicine: tailoring treatments based on individual genetic makeup.
- Cancer genomics: detecting somatic mutations and understanding tumor evolution.

Agriculture:

- Improving crop yields, pest resistance, and disease resistance through genomic selection.
- Understanding plant and animal genetics to breed more resilient species.

Evolutionary Biology:

- Understanding the genetic relationships between species.
- Tracing the evolution of genes, populations, and species over time.

Microbial Genomics:

- Studying the genomes of bacteria, viruses, and other microorganisms to identify pathogenic strains, antibiotic resistance, and virulence factors.

Challenges in WGS and Assembly

- **Cost:** While sequencing costs have dramatically decreased over the past decade, whole genome sequencing is still expensive, especially for large or complex genomes.
- **Data Storage:** WGS generates a massive amount of data, requiring significant storage and computational power for analysis.
- **Complexity:** Assembling large, complex genomes (such as human or plant genomes) is still challenging due to repetitive regions and structural variants.

- **Errors:** Sequencing errors (e.g., base-calling errors, misassemblies) can occur, which is why multiple sequencing technologies are often used to improve accuracy.

Whole genome sequencing and assembly are fundamental techniques in modern genomics. They allow researchers to gain deep insights into the genetic makeup of organisms, uncovering both common and rare genetic variations. The combination of cutting-edge sequencing technologies and advanced bioinformatics tools has made it possible to sequence and assemble entire genomes with increasing accuracy and cost-effectiveness. However, challenges such as data storage, assembly complexity, and sequencing errors remain active areas of research.

GENOME ANNOTATION TOOLS

Genome annotation is a crucial step in the post-sequencing process, where raw DNA sequence data is interpreted to identify functional elements within the genome. This process allows scientists to assign biological meaning to the sequences by identifying genes, regulatory regions, non-coding elements, and other genomic features. Proper annotation transforms a mere string of nucleotides into a comprehensible map of the organism's genetic makeup, helping researchers uncover insights into gene function, disease mechanisms, evolutionary processes, and much more. The complexity of genome annotation varies greatly between organisms, with eukaryotic genomes generally presenting more challenges due to their larger size, complex gene structures, and presence of repetitive sequences. In contrast, prokaryotic genomes are usually smaller and simpler but still require accurate identification of functional elements. To meet these challenges, a wide range of genome annotation tools has been developed, each tailored to different types of data (e.g., DNA sequence, RNA-seq) and specific annotation needs (e.g., gene prediction, functional classification). These tools utilize various approaches, from traditional gene prediction algorithms that rely on known sequence patterns to advanced methods that incorporate RNA-seq data, homology searches, and machine learning. The ultimate goal of these tools is to provide a comprehensive and accurate map of the genome, making it possible to annotate genes, identify regulatory networks, predict protein functions, and explore relationships between genomic features. The availability of user-friendly annotation platforms has made these tools accessible not only to professional researchers but also to the broader scientific community, enabling new discoveries across fields like medicine, agriculture, and evolutionary biology. Genome annotation is the process of identifying and marking functional elements in a sequenced genome, such as genes, regulatory regions, non-coding regions, and other important features. The goal is to provide biological context to the raw sequence data. Various tools and software platforms are available for genome annotation, each with its specific strengths and capabilities.

MOST COMMONLY USED GENOME ANNOTATION TOOLS

Gene Annotation Tools

Gene annotation involves identifying the locations of genes within the genome, predicting their structure, and assigning biological functions.

Augustus

- **Purpose:** AUGUSTUS is one of the most widely used gene prediction tools for eukaryotic genomes.
- **Features:**
  - It predicts genes, exons, and other genomic features in eukaryotic genomes, both for known species and new ones.
  - Supports a wide variety of organisms (plants, fungi, animals, etc.).
  - Highly accurate gene prediction, especially with existing training data.
  - Integrates with other tools like **RNA-seq** data for improved annotation.
- **Website:** AUGUSTUS

Genemark

- **Purpose:** GeneMark is a gene prediction tool that works well for both prokaryotic and eukaryotic genomes.
- **Features:**
  - Offers tools for **prokaryotic** (GeneMarkS) and **eukaryotic** (GeneMark-ES) genome annotation.
  - Can work with low-coverage data and limited genomic resources.
  - Predicates gene structures and provides genome-wide predictions.
- **Website:** GeneMark

Maker

- **Purpose:** MAKER is an automated pipeline for genome annotation that integrates multiple gene prediction tools.
- **Features:**
  - Supports **RNA-seq** and **EST** data for gene prediction.
  - Integrates results from **AUGUSTUS**, **GeneMark**, and **SNAP** for comprehensive annotation.
  - Useful for both eukaryotic and prokaryotic genomes.
  - Easy-to-use with web interfaces for analysis and result visualization.
- **Website:** MAKER

Functional Annotation Tools

These tools help to predict the function of genes, proteins, and other genomic features based on sequence homology to known databases.

BLAST (Basic Local Alignment Search Tool)

- **Purpose:** BLAST is a fundamental tool for sequence similarity search and functional annotation.
- **Key Features:**
  - Compares input sequences (nucleotide or protein) against databases like GenBank or UniProt to identify homologous sequences.
  - Useful for functional annotation by identifying conserved domains and possible functions based on sequence similarity.
  - Can be used for genome-wide annotation and identifying genes involved in specific pathways or functions.
  - **Website:** BLAST

InterProScan

- **Purpose:** InterProScan is a tool for the identification of protein domains, families, and functional sites.
- **Features:**
  - Combines multiple protein domain databases (e.g., Pfam, SMART, ProDom) to predict the structure and function of proteins.
  - Useful for annotating gene products with functional information based on known protein domains.
  - Offers detailed domain information and pathway mapping.
- **Website:** InterProScan

Gene Ontology (GO) Tools

- **Purpose:** GO provides a standardized system to annotate and classify genes based on their functions (biological processes, molecular functions, and cellular components).
- **Features:**
  - Tools like **AmiGO** allow users to visualize and interpret GO terms and their relationships.
  - **GOrilla** and **topGO** can be used to analyze over-representation of GO terms in different gene sets.

- GO annotations are used to classify genes into biological processes and functional pathways.

◉ **Websites:**

- AmiGO
- GOrilla
- Gene Ontology

RNA-SEQ BASED ANNOTATION TOOLS

RNA sequencing (RNA-seq) can be used to identify expressed genes, splice variants, and alternative splicing events. These tools are often used in conjunction with genome-wide data to improve gene annotation.

Cufflinks

◉ **Purpose:** Cufflinks is a tool designed to assemble and quantify transcriptomes from RNA-seq data.

◉ **Key Features:**

- Identifies novel transcript isoforms and alternative splicing events.
- Provides transcript-level gene expression quantification.
- Can be used to generate comprehensive gene annotations for both known and novel transcripts.
- **Website:** Cufflinks

StringTie

◉ **Purpose:** StringTie is a fast and highly efficient tool for transcript assembly and quantification.

◉ **Key Features:**

- Can assemble RNA-seq reads into full-length transcripts.
- Identifies gene structures and quantifies gene expression levels.
- Provides a suitable platform for annotating novel transcripts and performing differential expression analysis.

◉ **Website:** StringTie

Hisat2 + StringTie + Ballgown Pipeline

◉ **Purpose:** A comprehensive RNA-seq analysis pipeline combining **Hisat2** for read alignment, **StringTie** for transcript assembly, and **Ballgown** for visualization and analysis.

- **Key Features:**
  - Hisat2 performs accurate alignment of RNA-seq reads to the genome.
  - StringTie assembles transcripts and estimates gene expression levels.
  - Ballgown offers interactive visualization of expression data for differential analysis.
- **Website:** Hisat2, Ballgown

Prokaryotic Annotation Tools

For prokaryotic genomes, specialized tools are designed for predicting genes, operons, and regulatory regions.

RAST (Rapid Annotations using Subsystems Technology)

- **Purpose:** RAST is a widely used tool for annotating prokaryotic genomes.
- **Key Features:**
  - Provides automated functional annotation of microbial genomes.
  - Uses a curated subsystems-based approach to categorize genes based on their function in metabolic pathways.
  - Well-suited for bacteria, archaea, and other microorganisms.
- **Website:** RAST

Prokka

- **Purpose:** Prokka is a fast and reliable prokaryotic genome annotation tool.
- **Key Features:**
  - Provides rapid annotation of prokaryotic genomes with high accuracy.
  - Supports bacterial, archaeal, and plasmid genomes.
  - Uses databases like GenBank, RefSeq, and others for gene function predictions.
- **Website:** Prokka

Visualization Tools

Genome annotation often requires visualization to interpret results effectively.

JBrowse

- **Purpose:** JBrowse is a fast and user-friendly genome browser that allows visualization of genomic data.
- **Key Features:**

- Supports browsing annotations (genes, transcripts, SNPs) and other genomic features.
- Allows integration of various genomic data types such as RNA-seq, ChIP-seq, and variant data.
- Interactive and scalable for large genomes.

⦿ **Website:** JBrowse

IGV (Integrative Genomics Viewer)

⦿ **Purpose:** IGV is an interactive, high-performance visualization tool for browsing large-scale genomic data.

⦿ **Key Features:**

- Supports various data types such as genome sequences, variant calls, RNA-seq, and epigenomic data.
- Useful for visualizing annotated genes and RNA-seq results.
- Can handle large datasets with high performance.

⦿ **Website:** IGV

Genome annotation is an essential step in interpreting a newly sequenced genome. The tools mentioned above cover a wide range of applications, from gene prediction and functional annotation to RNA-seq-based improvements and visualization. Choosing the right tool depends on the organism, the quality of the sequence, and the specific goals of the annotation project. Many modern annotation pipelines integrate multiple tools to provide the most comprehensive results.

GENOME PREDICTION ALGORITHMS

Genome prediction algorithms are computational tools designed to analyze raw genomic sequence data and predict the location and structure of functional elements within a genome. These algorithms are essential for transforming raw DNA sequences into biologically meaningful information, such as identifying genes, regulatory regions, and other functional genomic features. The process of genome prediction involves various tasks, including gene finding, promoter identification, exon-intron boundaries, and predicting protein-coding regions, among others. Genome prediction is a fundamental step in genome annotation, and its complexity depends largely on the type of organism and the quality of the available data. In eukaryotic genomes, with their complex gene structures, alternative splicing, and large non-coding regions, prediction algorithms must contend with challenges like repetitive sequences, high gene density, and large genomic sizes. In contrast, prokaryotic genomes, while generally smaller and less complex, still require accurate predictions to capture important genomic features such as operons, regulatory elements, and mobile genetic elements. To address these challenges, a variety of genome prediction algorithms have been developed, ranging from traditional sequence-based methods to more advanced approaches that integrate machine learning, RNA-seq data, and comparative genomics. These algorithms employ different strategies—such as hidden Markov models (HMMs), dynamic programming, and graph-based methods—to predict the presence and structure of genes and other genomic features with varying levels of accuracy. As sequencing technologies advance, new algorithms continue to emerge, improving our ability to predict genomic features more accurately and efficiently, even for poorly characterized genomes.

Genome prediction algorithms are essential tools used in computational biology and bioinformatics for identifying and annotating genes and other functional regions within a genome. These algorithms aim to predict the locations of coding and non-coding regions, such as genes, promoters, introns, exons, and regulatory elements, based on the sequence data generated by high-throughput sequencing technologies.

There are several types of genome prediction algorithms, each designed to handle different challenges inherent in genome analysis, such as sequence complexity, gene density, and the presence of repetitive regions. These algorithms employ various techniques ranging from simple sequence-based methods to more sophisticated statistical models and machine learning approaches.

Types of Genome Prediction Algorithms

Gene Prediction Algorithms

Gene prediction algorithms are specifically designed to predict the locations of genes within a genome. This involves identifying start and stop codons, splice sites (exon-intron boundaries), and other important structural features of genes.

Ab Initio Gene Prediction Algorithms

Ab initio gene prediction algorithms make predictions based solely on the DNA sequence, without relying on experimental data like transcriptomic or proteomic information. These methods are typically based on statistical models that identify patterns in sequence data that are characteristic of coding regions and gene boundaries.

- **Hidden Markov Models (HMMs)**: One of the most widely used methods for gene prediction. HMMs are statistical models that represent the sequence data as a series of states (e.g., exon, intron, intergenic region), with transitions between these states defined by probabilities. Gene prediction algorithms like **GENEMARK**, **AUGUSTUS**, and **SNAP** use HMMs to predict the likelihood of each state at a given position in the genome, allowing the identification of gene structures. These models are trained on known genomes, and once trained, they can predict gene locations in novel genomes.

- **Example: AUGUSTUS** is one of the most successful gene prediction tools based on HMMs, capable of predicting genes in both eukaryotic and prokaryotic genomes.

◉ **Dynamic Programming Algorithms:** These algorithms are often used to identify exon-intron boundaries. The algorithm searches for the best alignment between the input sequence and a database of known gene structures. It uses a scoring system to reward alignments that match known patterns, such as conserved splice sites.

- **Example: GeneMark** uses dynamic programming for gene prediction, allowing it to handle genomes with varying levels of sequence complexity.

Evidence-based Gene Prediction Algorithms

Evidence-based gene prediction combines ab initio methods with experimental data, such as **RNA-seq** data or **EST** (Expressed Sequence Tag) data, to improve accuracy. These tools can take advantage of known genes from closely related organisms and incorporate additional layers of evidence to refine predictions.

◉ **Genewise:** A tool that refines predictions by aligning **RNA-seq** or **EST** sequences to a genome. It incorporates both sequence similarity and splice site predictions.

◉ **Maker:** An automated annotation pipeline that integrates ab initio gene predictions with evidence from RNA-seq and EST data. It uses multiple gene prediction algorithms like AUGUSTUS and GeneMark to improve the accuracy of gene predictions, especially for poorly annotated genomes.

Structural Genome Prediction Algorithms

These algorithms predict not only genes but also the structure of the genome, including the location of non-coding elements like regulatory regions (e.g., promoters, enhancers), operons, and transposable elements.

Promoter Prediction

Promoters are regions of DNA that initiate transcription of a gene. Predicting promoter regions is essential for understanding gene regulation.

◉ **PromoterHunter:** A tool that uses sequence motifs (such as TATA boxes and CpG islands) to predict promoter regions in eukaryotic genomes.

- **BPROM:** A software that uses position weight matrices (PWM) to predict bacterial promoters by identifying conserved sequence motifs upstream of genes.

Operon Prediction

Operons are clusters of genes in prokaryotic genomes that are transcribed together. Predicting operons is vital for understanding bacterial gene regulation.

- **DOOR:** A database of predicted operons in bacterial genomes, along with a tool for operon prediction using sequence motifs and known regulatory elements.

Non-Coding RNA (ncRNA) Prediction

Non-coding RNAs play critical roles in cellular processes, but they do not translate into proteins. Predicting ncRNAs such as microRNAs, tRNAs, and rRNAs is a key challenge.

- **tRNAscan-SE:** A tool for identifying tRNAs in prokaryotic and eukaryotic genomes based on sequence and structural motifs.
- **Infernal:** A program that uses covariance models to identify other types of ncRNAs, such as miRNAs, rRNAs, and snoRNAs, by aligning them to known RNA families.

Comparative Genome Prediction Algorithms

Comparative genomics involves comparing the genome of an organism to a reference genome or a set of closely related genomes. These algorithms are useful for transferring gene annotations between species and for predicting conserved genomic features across species.

Homology-based Prediction

Homology-based algorithms use sequence alignment to find regions of similarity between the query genome and a reference genome. If the sequence matches a known gene or genomic feature in a reference, it is predicted to have the same function.

- **BLAST (Basic Local Alignment Search Tool):** A fundamental sequence comparison tool used to align a query sequence to a reference database. It is widely used to identify homologous genes and predict their functions based on sequence similarity.
- **Exonerate:** A tool for aligning DNA or protein sequences to a reference genome. It is particularly useful for transferring gene annotations from a reference genome to a related organism.

Synergy-based Prediction

In some cases, algorithms may combine multiple types of evidence, including RNA-seq, DNA-seq, and protein sequences, to improve the prediction of genomic features.

- **MUMmer:** A tool used to align complete genomes and identify conserved regions, which can help predict gene synteny (conserved order of genes) between species.
- **LastZ:** A tool used for long DNA sequence alignments, often used for aligning whole-genome sequences and identifying conserved regions for gene prediction.

Machine Learning-Based Genome Prediction Algorithms

Machine learning approaches have recently gained popularity for genome prediction due to their ability to model complex relationships in large datasets. These algorithms use a training dataset of annotated genomes to learn patterns in sequence data that are indicative of various genomic features.

- **DeepGene:** A deep learning-based framework for predicting gene structures, including exon-intron boundaries, using deep neural networks. It has shown significant improvements over traditional methods, particularly in complex and poorly annotated genomes.
- **GenomeANNIS:** A machine learning tool that uses annotated genome data to predict gene structures. It employs various machine learning models, including support vector machines (SVMs) and decision trees.

Challenges in Genome Prediction

- **Repetitive Sequences:** Repetitive DNA sequences (e.g., transposons, tandem repeats) pose a challenge for many gene prediction algorithms. These sequences can lead to incorrect predictions of gene boundaries and gene copy numbers.
- **Incomplete Genomes:** Low-quality or incomplete sequencing can lead to gaps in the data, making it difficult for gene prediction algorithms to identify all of the genomic features accurately.
- **Complex Gene Structures:** Eukaryotic genomes, in particular, may contain complex gene structures, such as alternative splicing, overlapping genes, and multiple promoters, which can be difficult to predict accurately.
- **Species-Specific Features:** Certain species may have unique genomic features that require specialized prediction algorithms. For example, operon structure in bacteria or polycistronic genes in some plants require specific approaches.

Genome prediction algorithms are a crucial component of modern genomics and bioinformatics, providing the necessary tools to interpret the vast amounts of sequence data generated by high-throughput sequencing technologies. With advancements in statistical modeling, machine learning, and RNA-seq integration, these algorithms have significantly improved our ability to predict genes and other functional genomic features with greater accuracy. However, challenges remain in handling complex, incomplete, or poorly annotated genomes, and ongoing research continues to improve these algorithms to keep pace with the rapidly evolving field of genomics.

8 Chapter

RNA-SEQ AND TRANSCRIPTOMICS

INTRODUCTION

RNA sequencing (RNA-Seq) and transcriptomics are pivotal tools in bioinformatics for understanding gene expression and the functional complexity of genomes. RNA-Seq involves high-throughput sequencing technologies to capture and quantify RNA molecules, providing a comprehensive snapshot of the transcriptome—the complete set of RNA transcripts produced by a cell or tissue at a specific time. This method allows researchers to examine not only gene expression levels but also alternative splicing events, gene fusion events, and the presence of novel transcripts, offering insights into cellular processes and disease mechanisms. Transcriptomics, the broader field encompassing RNA-Seq, analyzes the structure, function, and dynamics of transcripts, enabling the identification of key biomarkers and regulatory pathways. Together, RNA-Seq and transcriptomics have revolutionized genomics research, facilitating advancements in personalized medicine, disease diagnosis, and therapeutic development. RNA-Seq (RNA Sequencing) and transcriptomics are essential fields in modern bioinformatics that enable the comprehensive study of gene expression and the complexities of cellular transcriptional activity. These techniques have transformed genomics by providing deeper insights into gene regulation, cellular function, and disease mechanisms.

RNA-SEQ

RNA-Seq is a high-throughput sequencing method used to analyze the transcriptome of a cell or organism. Unlike traditional techniques like microarrays, RNA-Seq doesn't rely on pre-designed probes or reference genomes. Instead, it sequences the entire population of RNA molecules in a sample, generating millions of short reads that correspond to the RNA present in the sample at the time of collection.

The RNA-Seq workflow typically includes the following steps:

- **RNA Extraction**: Total RNA is isolated from the tissue or cell sample.
- **Library Preparation**: RNA is converted into complementary DNA (cDNA) using reverse transcription. The cDNA is then fragmented, and adapters are ligated to the fragments to facilitate sequencing.

- **Sequencing**: The cDNA library is sequenced using a high-throughput sequencing platform (e.g., Illumina, Oxford Nanopore, or PacBio).
- **Data Analysis**: The raw sequence data is processed using bioinformatics tools to align the reads to a reference genome or transcriptome, quantifying gene expression levels and identifying novel transcripts, splicing variants, and other genomic features.

RNA-Seq allows for:

- **Quantitative gene expression**: It provides an accurate measurement of gene expression levels, expressed as the number of reads aligned to each gene or transcript.
- **Alternative splicing**: It reveals the various splicing isoforms of genes, helping researchers understand how genes are differentially spliced under different conditions.
- **Novel transcript discovery**: RNA-Seq can identify previously unknown transcripts, including non-coding RNAs (e.g., long non-coding RNAs, microRNAs).
- **Gene fusion detection**: RNA-Seq can identify chimeric gene transcripts formed by the fusion of two genes, which is crucial in cancer research.
- **Single-nucleotide resolution**: It can detect small-scale genetic variations like SNPs, indels, and RNA editing events that affect gene function.

Transcriptomics

Transcriptomics is the study of the transcriptome—the complete set of RNA molecules that are transcribed from the DNA of a cell or organism. While RNA-Seq is one of the primary techniques used in transcriptomics, the field also encompasses a wide range of approaches aimed at understanding the dynamic and intricate processes of gene expression.

Transcriptomics includes several key aspects:

- **Gene Expression Profiling**: Measuring the abundance of mRNA transcripts in different cells, tissues, or conditions to understand which genes are active and to what extent.
- **Alternative Splicing Analysis**: Understanding how genes can produce multiple protein isoforms through alternative splicing, a process that plays a significant role in cellular diversity and functional complexity.
- **Non-coding RNA Studies**: Investigating the roles of non-coding RNAs such as microRNAs, long non-coding RNAs (lncRNAs), and small nucleolar RNAs (snoRNAs), which regulate gene expression at various levels.

- **Differential Gene Expression (DGE):** Comparing gene expression between different conditions (e.g., normal vs. diseased tissues, treated vs. untreated samples) to identify genes involved in particular biological processes or disease states.
- **Gene Regulatory Networks:** Understanding how transcription factors, enhancers, and other regulatory elements control the expression of genes.

Transcriptomics can be applied to diverse biological questions:

- **Developmental Biology:** Studying changes in gene expression during development and differentiation.
- **Cancer Research:** Identifying gene expression signatures associated with different cancer types and stages.
- **Functional Genomics:** Understanding how genes contribute to cellular functions and their roles in metabolic pathways, signaling cascades, and disease processes.
- **Pharmacogenomics:** Investigating how gene expression profiles change in response to drug treatments.

Applications of RNA-Seq and Transcriptomics

- **Biomarker Discovery:** By identifying genes whose expression is altered in disease conditions, RNA-Seq helps in discovering potential biomarkers for diagnosis, prognosis, and therapeutic targets.
- **Cancer Genomics:** Transcriptomic analyses of cancer tissues reveal specific mutations, gene fusions, and differential gene expression profiles that can inform personalized treatment strategies.
- **Comparative Transcriptomics:** Comparing transcriptomes across different species, developmental stages, or environmental conditions to understand evolutionary changes or adaptational responses.
- **Functional Annotation of the Genome:** RNA-Seq data can be used to annotate the genome more accurately, identifying coding regions, untranslated regions (UTRs), and other regulatory elements.
- **Immune Profiling:** RNA-Seq can reveal gene expression patterns in immune cells, helping to understand immune responses in diseases like autoimmune disorders and infections.

Challenges in RNA-Seq and Transcriptomics

While RNA-Seq has revolutionized transcriptomic studies, there are several challenges that researchers face:

- **Data Complexity:** RNA-Seq generates vast amounts of data that require extensive computational resources for storage, processing, and analysis.
- **Normalization:** Accurate quantification of gene expression requires effective normalization to account for technical biases, such as differences in sequencing depth or RNA composition.
- **Transcript Assembly:** Mapping and assembling transcripts, particularly in organisms with complex genomes or limited reference data, can be challenging.
- **Interpretation of Non-coding RNAs:** While the role of protein-coding genes is well-understood, non-coding RNAs remain enigmatic, and their functional significance is still under active investigation.
- **Batch Effects:** Variability introduced by sample preparation, sequencing technology, and environmental conditions can confound results, making it essential to carefully design experiments and control for potential confounders.

RNA-Seq and transcriptomics are integral tools in bioinformatics and genomics, enabling researchers to delve into the complexity of gene expression and its regulation. These technologies have wide-ranging applications in biomedical research, from understanding the molecular basis of diseases to advancing precision medicine. Despite some technical challenges, the ability to capture and analyze the entire transcriptome provides unprecedented insights into cellular mechanisms, gene function, and genome regulation, making RNA-Seq a cornerstone of modern biological research.

BASICS OF RNA SEQUENCING

RNA sequencing (RNA-Seq) is a cutting-edge technology that allows researchers to analyze the entire transcriptome of an organism, providing comprehensive insights into gene expression, splicing events, transcript isoforms, and more. It has revolutionized the field of genomics by enabling the detection and quantification of RNA molecules in a highly detailed and quantitative manner. RNA-Seq has largely replaced older techniques like microarrays due to its ability to capture a broader range of transcript information, including novel and low-abundance transcripts.

Here's a detailed breakdown of RNA-Seq and its underlying concepts:

RNA Sequencing (RNA-Seq)

RNA-Seq is a high-throughput sequencing method that sequences the RNA present in a sample, typically converting the RNA into complementary DNA (cDNA) and then sequencing it using next-generation sequencing (NGS) technologies. The technique provides a snapshot of gene expression at a particular moment, showing which genes are active, their expression levels, and the structure of the RNA transcripts. Unlike other gene expression analysis methods like microarrays, RNA-Seq doesn't require a pre-defined set of probes and can capture the entire transcriptome, including known and novel RNA species, gene isoforms, and even non-coding RNAs (such as microRNAs and long non-coding RNAs).

STEPS INVOLVED IN RNA SEQUENCING

RNA-Seq involves several key steps, from sample preparation to data analysis. Here's an outline of the typical RNA-Seq workflow:

Step 1: RNA Extraction: The first step is to extract total RNA from the biological sample (e.g., cells, tissues, or organs). The quality and integrity of the RNA are critical for successful downstream analysis. The RNA is typically assessed using tools like the **Agilent Bioanalyzer** or **Nanodrop**, ensuring it is intact and free from contamination.

Step 2: mRNA Enrichment (Optional): For many RNA-Seq experiments, the focus is on **mRNA** rather than all RNA species. To enrich for mRNA, ribosomal RNA (rRNA) is often depleted, as rRNA makes up the majority of total RNA. This can be done using **rRNA removal kits** or by selecting mRNA based on the poly-A tail with magnetic beads coated with oligo(dT) sequences. However, if you want to profile the entire transcriptome, including non-coding RNAs, you may skip this enrichment step and sequence the total RNA directly.

Step 3: cDNA Library Preparation: Once the RNA has been isolated, it is reverse transcribed into **complementary DNA (cDNA)** using the enzyme **reverse transcriptase.** This cDNA serves as a more stable and amplifiable form of the RNA for sequencing.

- **Fragmentation:** The cDNA is fragmented into smaller pieces (typically 200-300 base pairs) to prepare for sequencing.
- **Adapter Ligation:** Sequencing adapters are ligated to the ends of the cDNA fragments. These adapters are short DNA sequences that help in the attachment of the cDNA fragments to the sequencing platform and allow for amplification and sequencing.

Step 4: Sequencing: The prepared cDNA library is loaded onto a **next-generation sequencing (NGS)** platform, such as **Illumina**, **PacBio**, or **Oxford Nanopore.** During sequencing, the cDNA fragments are read in parallel, generating millions of short sequence reads. The sequencing platforms work by reading the nucleotide sequence of each fragment (A, T, C, G) and generating a large set of raw sequencing data.

Step 5: Data Processing and Alignment: The raw sequence data (often in the form of FASTQ files) is processed to align the reads to a reference genome or transcriptome. Several tools are available for this step:

- **Trimming and Quality Control**: The raw sequencing data is processed to remove low-quality reads, adapter sequences, or any contaminating sequences.
- **Alignment**: The clean reads are mapped to a reference genome or transcriptome using alignment algorithms such as **STAR**, **TopHat2**, or **HISAT2**. These tools use algorithms to match the sequence reads to the reference genome, aligning them with known genes and transcript regions.
- **Quantification**: After alignment, the next step is to quantify gene expression by counting the number of reads aligned to each gene. This can be done using tools like **Cufflinks**, **HTSeq**, or **featureCounts**. These tools count reads corresponding to each gene or transcript to estimate their abundance.

Step 6: Differential Expression and Analysis: Once the data has been aligned and quantified, the next step is to identify differentially expressed genes (DEGs) between experimental conditions (e.g., disease vs. normal tissue, treated vs. untreated samples). Common tools for differential expression analysis include **DESeq2**, **edgeR**, and **limma**.

Additional analyses include:

- **Alternative splicing**: RNA-Seq data allows the identification of alternative splicing events (e.g., exon skipping, intron retention) within genes.
- **Gene fusion detection**: In certain contexts, such as cancer research, RNA-Seq can help identify gene fusions that may not be detectable through DNA sequencing alone.

Concepts in RNA-Seq

- **Gene Expression Quantification:** RNA-Seq provides highly accurate, quantitative data about gene expression. The number of reads mapped to a gene or transcript is proportional to its abundance in the sample. This allows researchers to compare the relative expression levels of different genes or transcripts between different samples, conditions, or time points.
- **Alternative Splicing:** RNA-Seq allows for the detection of **alternative splicing**, where a single gene can produce multiple distinct RNA transcripts (isoforms) due to different splicing patterns. This process greatly increases the diversity of proteins that can be produced from a single gene and is a key regulatory mechanism in eukaryotic cells.

- **Non-Coding RNAs:** RNA-Seq can also be used to profile non-coding RNAs (ncRNAs), which include **microRNAs, long non-coding RNAs (lncRNAs), small nuclear RNAs (snRNAs)**, and **small nucleolar RNAs (snoRNAs)**. These molecules do not code for proteins but play essential roles in gene regulation, chromatin remodeling, and RNA processing.
- **Single-Cell RNA-Seq:** RNA-Seq has evolved into **single-cell RNA sequencing (scRNA-Seq)**, which allows for the analysis of gene expression at the individual cell level. This technique reveals cellular heterogeneity in tissues and can identify rare cell types or subpopulations that would be masked in bulk RNA-Seq analyses.

Applications of RNA-Seq

RNA-Seq has broad applications in biological and biomedical research:

- **Gene Expression Profiling**: Comparing gene expression in different tissues, developmental stages, or experimental conditions.
- **Cancer Genomics:** Identifying tumor-specific gene expression profiles, gene fusions, mutations, and alternative splicing events in cancer.
- **Developmental Biology**: Studying the dynamic changes in gene expression during development, differentiation, and cellular responses to external stimuli.
- **Functional Genomics:** Understanding the function of unknown genes by correlating expression data with phenotypic outcomes.
- **Drug Development and Toxicology**: Identifying potential drug targets by profiling gene expression in response to different drug treatments or environmental stresses.
- **Personalized Medicine**: Tailoring medical treatments based on individual gene expression profiles.

Advantages of RNA-Seq

- **Comprehensive**: RNA-Seq provides a comprehensive view of the transcriptome, including all RNA species, known and unknown.
- **High Sensitivity**: RNA-Seq is highly sensitive, detecting even low-abundance transcripts that are often missed by microarrays.
- **Quantitative**: It allows for quantitative gene expression analysis without the need for standardization through pre-designed probes.
- **Novel Transcript Discovery**: RNA-Seq can identify novel transcripts, including new isoforms, non-coding RNAs, and gene fusions.

- **Resolution:** It provides high-resolution data, detecting alternative splicing, RNA editing, and other complex transcriptome events.

Challenges and Limitations of RNA-Seq

- **Cost:** RNA-Seq can be expensive, especially for large-scale or high-depth studies.
- **Data Analysis:** The bioinformatics tools required for RNA-Seq analysis can be complex, requiring significant computational power and expertise.
- **RNA Quality:** RNA-Seq requires high-quality RNA, and degradation or contamination of RNA can significantly impact the results.
- **Interpretation:** While RNA-Seq provides a wealth of information, interpreting complex datasets, especially with non-model organisms or unknown transcripts, can be challenging.
- RNA-Seq has emerged as a transformative tool in genomics, offering a powerful and versatile approach to studying gene expression and the transcriptome. Its ability to provide deep, quantitative, and comprehensive insights into RNA biology makes it invaluable in diverse fields, from basic research to clinical applications like cancer genomics and personalized medicine. While the technique comes with some challenges, such as cost and computational complexity, its advantages in sensitivity, resolution, and the ability to detect novel transcripts far outweigh these limitations, making RNA-Seq an indispensable tool in the modern molecular biologist's toolkit.

TRANSCRIPTOMIC DATA ANALYSIS

Transcriptomic data analysis involves the interpretation of data generated from high-throughput sequencing techniques, most commonly RNA sequencing (RNA-Seq), to explore the full range of RNA molecules present in a cell or tissue at a given time. The primary goal of transcriptomic data analysis is to understand the expression levels of genes, discover novel transcripts, identify alternative splicing events, and uncover regulatory mechanisms that control gene expression. Given the complexities involved in RNA-Seq data, a systematic and multi-step approach is required to accurately process and interpret the results. In this detailed explanation, we will go through the steps involved in transcriptomic data analysis, the various types of data generated, and the bioinformatics tools and approaches used at each stage.

Preprocessing of Raw RNA-Seq Data

The first step in transcriptomic data analysis is to process the raw sequencing data (typically in **FASTQ** format) produced by high-throughput sequencing platforms (like **Illumina**, **PacBio**, or **Oxford Nanopore**). These data can contain various sources of errors and artifacts that need to be addressed before any meaningful analysis can take place.

Quality Control (QC) and Filtering

Before proceeding with alignment and further analysis, the quality of raw sequencing reads must be assessed and improved. Common QC steps include:

- **Trimming:** Removing adapter sequences and low-quality bases from the ends of reads. Tools like **Cutadapt** and **Trimmomatic** are used for this task.
- **Filtering:** Discarding reads with low-quality scores (e.g., Phred score < 20) or too short in length (e.g., less than 50 bases).
- **Visualizing Quality:** Using tools like **FastQC** to generate reports and plots to assess the overall quality of the sequencing data.

By applying these quality control methods, you can ensure that the data used for downstream analysis are reliable and represent the biological sample as accurately as possible.

Read Alignment

After quality control, the next step is to **align** the cleaned reads to a reference genome or transcriptome. This step is essential to map the sequencing reads to known genomic locations, allowing for the identification of genes, transcripts, and alternative splicing events. The alignment process also helps quantify gene expression.

Alignment Tools

There are several popular algorithms and software packages for read alignment, including:

- **STAR:** Highly efficient and fast, STAR is used to align reads to a reference genome or transcriptome, and it can handle spliced alignments (i.e., when reads span exon-exon junctions).
- **HISAT2:** A fast and memory-efficient tool designed to align reads to large genomes, like the human genome. HISAT2 can handle splicing, multiple alignments, and mismatches.
- **TopHat2:** An older tool, but still used in some workflows for spliced read alignment, particularly with the **Bowtie** aligner.

Alignment produces **SAM** or **BAM** files, which store the mapped positions of the reads on the reference genome or transcript.

Transcript Assembly and Quantification

Once the reads are aligned, the next step is to **quantify gene expression** and **assemble transcripts**. Transcript quantification involves determining how many reads map to each gene, and transcript assembly helps to discover novel or previously unannotated transcripts.

Transcript Assembly

In RNA-Seq analysis, transcript assembly refers to the process of reconstructing full-length transcripts from fragmented reads. This is particularly useful when working with non-model organisms or when trying to identify novel transcripts. Common tools for transcript assembly include:

- **Cufflinks:** Used to assemble aligned reads into transcripts and estimate gene expression levels. It also enables the detection of alternative splicing events.
- **StringTie:** Similar to Cufflinks, but faster and more efficient. It assembles transcripts and quantifies gene expression levels.
- **Trinity:** Specifically designed for **de novo** transcriptome assembly (i.e., when no reference genome is available).

Quantification of Gene Expression

After assembling the transcripts or aligning directly to known gene models, the next task is to measure the expression of individual genes or transcripts. This step is typically done by counting the number of reads that map to each gene or transcript, often normalized to account for sequencing depth or gene length. Common tools for quantification include:

- **HTSeq:** A Python package that counts the number of reads mapped to each gene or feature based on an annotated genome.
- **featureCounts:** A fast tool for counting reads mapped to genomic features (e.g., genes, exons) and can be used with different aligners.
- **RSEM (RNA-Seq by Expectation Maximization):** A tool for transcript-level quantification that allows for the estimation of the abundance of each transcript in RNA-Seq data.

Normalization of read counts is often necessary to account for differences in sequencing depth across samples and genes' inherent lengths. **TPM (Transcripts per Million)**, **RPKM (Reads per Kilobase per Million)**, and **FPKM (Fragments per Kilobase per Million)** are common normalization methods.

Differential Expression Analysis

Once gene expression levels are quantified, the next step is to compare gene expression across different experimental conditions (e.g., treatment vs. control, disease vs. healthy). The goal of differential expression analysis is to identify genes that are significantly upregulated or downregulated in response to specific conditions.

Statistical Tools for Differential Expression

There are several bioinformatics tools designed to identify **differentially expressed genes (DEGs)**:

- **DESeq2:** A widely used tool based on a statistical model that uses the negative binomial distribution to identify DEGs. It normalizes data and performs statistical testing to compare gene expression across conditions.
- **edgeR:** Similar to DESeq2, edgeR uses a negative binomial distribution for RNA-Seq data. It is particularly suited for small sample sizes and low count data.
- **limma:** A linear modeling approach that is used in conjunction with **voom,** a method for modeling RNA-Seq data as if it were microarray data, allowing for differential expression analysis.

These tools calculate the **fold change** in gene expression and assign **p-values** to test the null hypothesis (no difference between conditions). To adjust for multiple comparisons, a **false discovery rate (FDR)** correction is often applied using methods such as **Benjamini-Hochberg.**

Gene Ontology (GO) and Pathway Enrichment Analysis

After identifying DEGs, researchers often perform **functional enrichment analysis** to understand the biological significance of the differentially expressed genes. This analysis provides insights into biological processes, molecular functions, and cellular components that are over-represented in the data.

Gene Ontology (GO) Analysis

GO analysis categorizes genes based on their roles in biological processes (e.g., cell division, response to stress), molecular functions (e.g., enzyme activity, binding), and cellular components (e.g., nucleus, mitochondria). Popular tools for GO enrichment analysis include:

- **DAVID** (Database for Annotation, Visualization, and Integrated Discovery)
- **ClusterProfiler:** A tool for R that provides an easy interface to perform GO analysis and visualization of enrichment results.

Pathway Enrichment Analysis

Gene set enrichment analysis (GSEA) can help identify specific **biological pathways** associated with DEGs. Pathway databases like **KEGG** (Kyoto Encyclopedia of Genes and Genomes), **Reactome**, and **WikiPathways** provide curated lists of genes involved in metabolic and signaling pathways. Tools like **GSEA** and **ReactomePA** can identify which pathways are significantly enriched in your dataset.

Alternative Splicing and Transcript Isoforms

RNA-Seq is particularly useful for detecting **alternative splicing** events, where different exons are included or excluded in the final transcript. These events contribute to proteomic diversity and are critical in many biological processes.

Tools for Alternative Splicing Detection

- **DEXSeq:** A package designed to detect differential exon usage across different conditions.
- **rMATS** (RNA-Seq by MATS): A tool for identifying and quantifying alternative splicing events, such as exon skipping, alternative 3'/5' splice sites, and intron retention.
- **SpliceSeq:** An approach for analyzing RNA-Seq data to detect and visualize alternative splicing events.

Visualization of Transcriptomic Data

Data visualization plays a key role in interpreting and presenting the results of transcriptomic analyses. Some common visualization methods include:

- **Volcano Plots:** Used to display the significance (p-value) versus the magnitude of fold changes of differentially expressed genes.
- **Heatmaps:** Useful for showing the expression patterns of DEGs across samples or conditions.
- **PCA (Principal Component Analysis):** Used to reduce the dimensionality of RNA-Seq data and visualize sample clustering based on gene expression.
- **Gene Expression Plots:** Such as **boxplots**, **MA plots**, and **bar graphs**, to show gene expression across different conditions.

Transcriptomic data analysis is a complex and multi-step process that transforms raw RNA-Seq data into biologically meaningful insights. From quality control to alignment, quantification, differential expression analysis, and functional interpretation, each step requires careful attention to detail. The tools and methods mentioned above are essential for effectively analyzing and interpreting transcriptomic data. Through these analyses, researchers can uncover the intricate mechanisms that regulate gene expression, identify biomarkers, and provide new insights into disease biology, drug response, and cellular functions. As RNA-Seq technology continues to evolve, the scope of transcriptomic data.

DIFFERENTIAL GENE EXPRESSION ANALYSIS

Differential gene expression analysis (DGE) is one of the core applications of transcriptomic research, aimed at identifying genes whose expression levels significantly differ across experimental conditions, such as disease vs. healthy states, treated vs. untreated samples, or different developmental stages. The goal of DGE analysis is to determine which genes are upregulated or downregulated under specific biological conditions, shedding light on molecular mechanisms that drive cellular processes, disease progression, or responses to treatment. By comparing gene expression between different conditions, researchers can identify genes that play pivotal roles in the phenotype being studied, making DGE a fundamental tool in understanding biological variability and uncovering biomarkers for disease diagnosis, prognosis, and therapeutic intervention.

The Basis of Differential Expression

At the heart of differential gene expression analysis lies the comparison of gene expression levels between two or more conditions. Gene expression levels are typically quantified as the number of sequencing reads that map to a particular gene or transcript. This count is proportional to the amount of mRNA present in the sample and, consequently, the level of gene expression. Differential expression is based on statistical comparisons between the expression levels of genes across different groups or conditions. These comparisons aim to detect genes whose expression is significantly altered in response to environmental stimuli, disease, or other biological processes. The first step in DGE analysis is to accurately measure gene expression. With RNA-Seq, this involves mapping the raw sequencing reads to a reference genome or transcriptome to identify which genes the reads come from. The number of reads mapped to each gene is counted, and this count is used to estimate the gene's expression level. However, because RNA-Seq data can be influenced by factors such as sequencing depth, gene length, and sample variability, these raw counts must be normalized before any meaningful comparisons can be made.

Normalization in Differential Expression Analysis

Normalization is an essential step in RNA-Seq data analysis because it accounts for biases and technical factors that can affect gene expression measurement. One of the main challenges in RNA-Seq is the inherent differences in sequencing depth (i.e., the total number of reads generated in each sample), gene length, and RNA composition between samples. Without proper normalization, these differences can lead to misleading conclusions.

Common normalization methods include:

- **Total Count Normalization:** This method scales the raw read counts by the total number of reads in a sample, assuming that most genes are not differentially expressed between samples. This approach, while simple, can be biased when a small fraction of genes is highly expressed in one condition.
- **Reads Per Kilobase per Million (RPKM):** This method normalizes for both the sequencing depth and the gene length. RPKM values are calculated by dividing the raw read count for a gene by the total number of reads and the gene length in kilobases. This adjustment helps compare the expression levels of genes of different sizes.
- **Fragments Per Kilobase per Million (FPKM):** A similar method to RPKM but used when paired-end sequencing is performed. It adjusts for both the gene length and sequencing depth, taking into account the fact that paired-end reads represent two parts of a single transcript.
- **Transcripts Per Million (TPM):** Unlike RPKM and FPKM, TPM normalizes first by the length of the gene and then by the total number of reads in the sample. This method has the advantage of being directly comparable across samples because the total sum of TPM values in each sample is the same.
- **Quantile Normalization:** In some cases, quantile normalization is used across samples to ensure that the distribution of expression levels is consistent. This is particularly helpful when comparing conditions where technical variation may obscure biological differences.

Once the data is normalized, researchers can proceed with identifying genes whose expression differs significantly between experimental conditions.

STATISTICAL METHODS FOR DIFFERENTIAL EXPRESSION

The next step is the **statistical analysis** of gene expression differences to identify which genes show significant changes across conditions. Several statistical methods are used to detect differential expression, each with its strengths and limitations. The primary goal is to identify genes with a **statistically significant** change in expression, often after correcting for multiple testing to avoid false positives.

Some commonly used statistical methods for differential expression analysis include:

- **T-tests/ANOVA:** These classical statistical methods compare the means of gene expression levels between two or more groups. While they are simple to apply, they assume that the data is normally distributed and homoscedastic (i.e., the variance is the same across groups), which is often not true for RNA-Seq data. This makes these methods less suitable for high-throughput data with count-based distributions.

- **Negative Binomial Distribution:** RNA-Seq data typically follow a **Poisson** or **negative binomial distribution** rather than a normal distribution. The negative binomial model accounts for overdispersed count data (i.e., where the variance is greater than the mean) and is the basis for many specialized tools for differential expression analysis. Tools like **DESeq2**, **edgeR**, and **limma** use this distribution to model gene expression and identify DEGs. These tools perform statistical testing using likelihood ratio tests or quasi-likelihood tests to determine whether the observed differences are statistically significant.
- **DESeq2:** One of the most popular and widely used tools for differential gene expression analysis, DESeq2 uses a negative binomial distribution to model RNA-Seq count data. It applies a **shrinkage estimation** method to improve the accuracy of fold change estimates, especially for lowly expressed genes. DESeq2 also offers methods for normalizing data, correcting for batch effects, and visualizing results. A key feature of DESeq2 is its ability to estimate size factors for each sample to adjust for differences in sequencing depth.
- **edgeR:** Another widely used tool, edgeR also models RNA-Seq data using the negative binomial distribution. It provides robust methods for testing differential expression in both small and large datasets. edgeR uses **empirical Bayes estimation** to account for biological variability, and it can be used with both count data and normalized data for various types of RNA-Seq experiments. It also supports multiple testing corrections to account for the large number of tests performed.
- **limma with voom:** The **limma** package, originally developed for microarray data, has been adapted for RNA-Seq data analysis using the **voom** method. Voom transforms RNA-Seq count data to log2 counts per million (logCPM) and estimates the mean-variance relationship. This allows for the use of linear modeling and empirical Bayes techniques to detect differentially expressed genes, even with small sample sizes. Limma with voom is a popular choice when RNA-Seq data contains complex experimental designs and multiple comparisons.

After performing statistical tests, a **p-value** is calculated for each gene to assess whether the difference in expression between conditions is statistically significant. A **false discovery rate (FDR)** adjustment is commonly applied (e.g., **Benjamini-Hochberg correction**) to account for the fact that when testing thousands of genes, some results will appear significant by chance alone. Genes with a **FDR-**

adjusted p-value below a predefined threshold (typically < 0.05) are considered differentially expressed.

Interpretation of Differential Expression Results

Once significant differentially expressed genes (DEGs) are identified, researchers need to interpret the results in the context of the biological system being studied. The direction of gene expression (upregulated or downregulated) can provide valuable insights into the underlying biological processes. For instance, upregulation of genes involved in inflammation or immune response may indicate an immune reaction, while downregulation of genes related to cellular metabolism could suggest metabolic dysregulation.After identifying DEGs, researchers typically perform **functional annotation** to explore the roles of these genes within biological pathways and cellular processes. **Gene Ontology (GO)** enrichment analysis is often performed to identify significant biological processes, molecular functions, and cellular components associated with the DEGs. This can help categorize genes into functional groups and provide insights into the mechanisms of disease or treatment response. Additionally, **pathway enrichment analysis** is frequently done using tools such as **KEGG** (Kyoto Encyclopedia of Genes and Genomes), **Reactome**, or **Gene Set Enrichment Analysis (GSEA)** to identify pathways that are significantly enriched with differentially expressed genes. Understanding which pathways are activated or suppressed can reveal how biological systems adapt to changes in environmental conditions, disease states, or therapeutic interventions.

Visualization of Differential Expression Results

Visualization is an important part of differential gene expression analysis. Key plots and charts are often used to summarize the results and communicate the findings clearly. Some commonly used visualizations include:

- **Volcano Plots:** A scatter plot that shows the relationship between the fold change (on the x-axis) and the statistical significance (on the y-axis) of each gene. Genes with large fold changes and high statistical significance (low p-values) appear at the top corners of the plot and are considered the most interesting.
- **Heatmaps:** Heatmaps are used to visualize the expression patterns of a set of differentially expressed genes across different samples or conditions. The heatmap colors represent the intensity of expression, making it easy to spot trends and clusters of genes with similar expression profiles.
- **Principal Component Analysis (PCA):** PCA is used to reduce the dimensionality of gene expression data and visualize how different samples cluster based on their gene expression profiles. PCA is useful for checking the quality of the data, identifying batch effects, and visualizing the biological variation in the dataset.

- **Gene Expression Plots:** Bar plots, boxplots, and MA plots are commonly used to illustrate changes in gene expression for individual genes across different conditions. These plots are useful for visualizing the magnitude and direction of expression changes in the most significant genes.

Chapter 9

ALGORITHMS IN BIOINFORMATICS

INTRODUCTION

Algorithms in bioinformatics play a pivotal role in analyzing and interpreting biological data, particularly in genomics, proteomics, and systems biology. As the volume and complexity of biological information grow exponentially, efficient algorithms are essential for processing large datasets, identifying patterns, and making meaningful biological inferences. These algorithms enable tasks such as sequence alignment, protein structure prediction, gene expression analysis, and phylogenetic tree construction, among others. By integrating computational techniques with biological research, bioinformatics algorithms help accelerate discoveries in fields like drug design, personalized medicine, and evolutionary biology, thereby advancing our understanding of life at the molecular level.

STRING MATCHING AND ALIGNMENT ALGORITHMS

String matching and alignment algorithms are fundamental tools in bioinformatics, primarily used for comparing biological sequences such as DNA, RNA, and protein sequences. These algorithms help identify similarities, mutations, and evolutionary relationships between sequences, which are crucial for tasks like gene identification, functional annotation, and phylogenetic analysis. String matching refers to the process of finding exact or approximate matches of a short string (or pattern) within a longer string. In bioinformatics, this often involves searching for specific motifs or patterns within large genetic sequences. The most well-known algorithms for string matching include the Knuth-Morris-Pratt (KMP), Rabin-Karp, and Boyer-Moore algorithms, which aim to efficiently locate patterns within a sequence by reducing the number of comparisons. On the other hand, sequence alignment involves arranging two or more sequences in a way that maximizes their similarities while accounting for insertions, deletions, and substitutions (mutations) that may have occurred over time. There are two primary types of sequence alignment: global alignment and local alignment.

- Global alignment algorithms, such as Needleman-Wunsch, aim to align entire sequences from end to end, making them suitable for comparing sequences of similar length and overall homology.

- Local alignment algorithms, such as Smith-Waterman, focus on finding the best matching subsequences within larger sequences, allowing for more flexible comparisons when sequences have varying lengths or contain highly divergent regions.

These alignment techniques form the backbone of many bioinformatics tools and databases, such as BLAST and ClustalW, and are essential for tasks ranging from gene finding and functional annotation to protein structure prediction and evolutionary analysis. String matching and alignment algorithms continue to evolve, driven by the growing complexity and size of biological datasets, as well as the need for faster and more accurate methods in fields like personalized medicine and genomics.

DYNAMIC PROGRAMMING AND GREEDY ALGORITHMS

In bioinformatics, both dynamic programming (DP) and greedy algorithms are essential techniques used to solve a variety of computational problems, especially those involving sequences, optimization, and combinatorial search problems. These algorithms are commonly applied to tasks like sequence alignment, motif finding, RNA secondary structure prediction, and protein folding. Below is a detailed description of both types of algorithms and their applications in bioinformatics.

DYNAMIC PROGRAMMING ALGORITHMS

Dynamic programming is a method used to solve complex problems by breaking them down into simpler subproblems. The key feature of DP is that it avoids redundant computation by storing the results of subproblems (often in a table or matrix) and reusing these results when needed. This technique is particularly useful when a problem has overlapping subproblems and optimal substructure — meaning that the optimal solution to the problem can be constructed efficiently from the optimal solutions to its subproblems.

Key Principles of Dynamic Programming

- **Optimal Substructure**: The problem can be solved by solving smaller subproblems, and the solution to the overall problem can be constructed from these subproblems.
- **Overlapping Subproblems**: The same subproblems appear multiple times in the process of solving the larger problem.

Applications in Bioinformatics

- **Sequence Alignment:** One of the most prominent applications of dynamic programming in bioinformatics is in **sequence alignment.** The Needleman-Wunsch algorithm for **global alignment** and the Smith-Waterman algorithm for **local alignment** are both based on dynamic programming principles. In these algorithms, a matrix is used to keep track of the optimal alignment score for every possible pair of positions in the two sequences being aligned. The DP table is filled in a bottom-up manner, with each cell representing the best possible alignment score for a pair of subsequences.

- **Needleman-Wunsch (Global Alignment)**: This algorithm aligns two sequences from start to finish, considering all possible insertions, deletions, and substitutions along the entire sequence. It works by filling a DP matrix where each cell represents the optimal score of aligning the first i characters of one sequence with the first j characters of another sequence.
- **Smith-Waterman (Local Alignment)**: This algorithm, also using dynamic programming, focuses on finding the best matching subsequence between two sequences, allowing for gaps or mismatches. It is especially useful when the sequences have regions of high similarity but are not globally similar.
- **RNA Secondary Structure Prediction:** RNA secondary structure prediction also relies on dynamic programming to predict the most stable secondary structure of an RNA molecule, considering base-pairing constraints and minimizing the total energy of the folded structure. Algorithms such as the **Nussinov algorithm** are widely used for this purpose.
- **Edit Distance**: Another classic DP problem in bioinformatics is calculating the **edit distance** (also known as Levenshtein distance) between two strings, which represents the minimum number of single-character edits (insertions, deletions, or substitutions) required to convert one string into another. This is often applied in DNA sequence comparison and variant detection.

Example: Needleman-Wunsch Algorithm

In global sequence alignment, the Needleman-Wunsch algorithm constructs a matrix where each cell contains the alignment score between two substrings of the sequences. It uses a recurrence relation based on three possibilities for each pair of sequence elements:

- **Match or mismatch**: The score is updated based on whether the characters match or differ.
- **Insertion or deletion (gap)**: The score accounts for the cost of inserting or deleting a character.

The optimal solution is found by backtracking through the matrix once it is filled.

GREEDY ALGORITHMS

Greedy algorithms are another approach to solving optimization problems, characterized by making the locally optimal choice at each stage with the hope of finding a global optimum. The main difference between greedy algorithms and dynamic programming is that greedy algorithms do not guarantee an optimal solution in all cases, but they tend to be more efficient in terms of time complexity.

Principles of Greedy Algorithms

- **Locally Optimal Choice:** At each step, the algorithm selects the best possible option available, without considering the global consequences of the decision.
- **No Backtracking:** Greedy algorithms do not revisit previous choices once made, which can sometimes lead to suboptimal solutions.

Applications in Bioinformatics

While greedy algorithms are less widely applicable to complex problems like sequence alignment, they are useful in other areas of bioinformatics that require fast, approximate solutions.

- **Multiple Sequence Alignment (MSA):** Greedy algorithms are used in methods such as **ClustalW** and **T-Coffee**, which progressively align sequences by adding one sequence at a time to a growing alignment. The algorithm greedily chooses the best alignment at each step based on pairwise comparisons, gradually building the alignment.
- **Motif Finding:** Greedy approaches can be used for finding conserved sequence motifs in a set of sequences, where the goal is to iteratively expand the current motif by adding one nucleotide or amino acid at a time based on local optimality.
- **Clustering and Phylogenetic Tree Construction:** Some algorithms for constructing phylogenetic trees, such as **Neighbor Joining** and **UPGMA**, use a greedy approach. At each step, the algorithm selects the two closest taxa (or clusters) to join together, building the tree by locally minimizing the distances between taxa.
- **Approximation Algorithms:** Greedy algorithms are also used in approximation problems, where finding an exact solution might be computationally expensive. For example, the **Traveling Salesman Problem** (TSP) can be approached with a greedy heuristic to find a good, though not necessarily optimal, route for visiting all cities.

Example: Greedy Multiple Sequence Alignment (MSA)

In **progressive MSA** methods like ClustalW, the algorithm first aligns the two most similar sequences. Then, it progressively adds the next most similar sequence to the growing alignment, greedily selecting pairs of sequences that maximize the alignment score at each step. This process continues until all sequences are aligned. While this method is fast and effective for many cases, it does not guarantee the globally optimal alignment.

Comparison of Dynamic Programming and Greedy Algorithms

Feature	Dynamic Programming	Greedy Algorithms
Optimality	Guarantees an optimal solution	Does not always guarantee an optimal solution
Complexity	Typically higher time and space complexity (e.g., O(n^2))	Generally faster (greedy choices can be made in linear time)
Memory Usage	Uses extra memory to store results of subproblems	Uses minimal memory, only keeping current states
Appropriate Problems	Problems with overlapping subproblems and optimal substructure (e.g., sequence alignment)	Problems where local optimal choices lead to global optimality (e.g., MSA, clustering)
Example Applications	Sequence alignment, RNA structure prediction, edit distance	Multiple sequence alignment, motif finding, clustering

Both dynamic programming and greedy algorithms have their strengths and are widely used in bioinformatics to solve different types of problems. Dynamic programming is essential for complex, optimization-based problems that require an exact solution and where overlapping subproblems are present, such as in sequence alignment or RNA structure prediction. On the other hand, greedy algorithms offer a faster, more efficient solution in situations where approximate solutions are acceptable, such as in multiple sequence alignment and clustering tasks. Understanding when and how to use these algorithms is crucial for tackling the diverse range of problems encountered in computational biology.

HIDDEN MARKOV MODELS (HMM)

A Hidden Markov Model (HMM) is a statistical model that represents a system which undergoes a series of states that are not directly observable (hence "hidden"). Instead, the system produces observable outputs (or emissions) that depend probabilistically on the hidden state it is in. HMMs are widely used in bioinformatics to model sequential data where the underlying process is not directly observable, but the data itself contains informative patterns. HMMs have been applied extensively in areas like sequence alignment, gene prediction, protein structure prediction, motif discovery, and RNA secondary structure prediction, among other tasks.

Components of a Hidden Markov Model

An HMM is formally defined by the following elements:

- **States:** The system is assumed to be in one of a set of hidden states at any given time, but we cannot observe the states directly. These states represent abstract biological concepts (e.g., regions of a gene, different secondary structure motifs in proteins). Each state has a probability distribution over the possible observable outputs (or emissions).
- **Observations:** The observations are the actual data points that we can observe (e.g., DNA sequences, protein sequences, or other biological data). In the case of bioinformatics, each observation typically corresponds to a symbol from an alphabet (e.g., nucleotide or amino acid) that is emitted by the hidden state.
- **Transition Probabilities:** These represent the probability of moving from one hidden state to another between time steps. In the context of biological sequences, these probabilities can represent the likelihood of moving from one type of motif or domain to another (e.g., from an exon to an intron in a gene).
- **Emission Probabilities:** Each hidden state is associated with an emission probability distribution, which represents the likelihood of observing a particular observation given the state. In gene prediction, for example, the emission probability could represent the likelihood of seeing a particular nucleotide at a given position in a coding or non-coding region of a gene.
- **Initial Probabilities:** These probabilities represent the likelihood that the system starts in each possible hidden state. In bioinformatics, this can correspond to the probability of a particular region of DNA starting with a gene or a promoter sequence.

How HMMs Work

The process of an HMM involves two main stages: the **hidden process** and the **observed process.**

- **Hidden Process:** The model begins in a hidden state according to the initial probabilities. From there, the state transitions to other states according to the transition probabilities, which might depend on the current state. The state transitions occur probabilistically, meaning that at each step, there is a fixed probability of moving to a particular state based on the current state.

- **Observed Process:** In each state, the model produces an observation (e.g., a nucleotide or amino acid) according to the emission probabilities of that state. This is the output that we can see and analyze. The sequence of observations is assumed to depend only on the hidden state and not on the history of past observations (this is the Markov property).

Mathematical Representation of an HMM

An HMM can be mathematically described as follows:

- Let the set of hidden states be denoted by S={s1,s2,...,sN}$S=\{s1,s2,\ldots,sN\}$.
- Let the set of possible observations be denoted by O={o1,o2,...,oM}$O=\{o1,o2,\ldots,oM\}$.
- The transition probability matrix A={aij}$A=\{aij\}$, where aijaij is the probability of transitioning from state sisi to state sjsj.
- The emission probability matrix B={bij}$B=\{bij\}$, where bijbij is the probability of observing ojoj in state sisi.
- The initial probability vector π={πi}$\pi=\{\pi i\}$, where πiπi is the probability of starting in state sisi.

Given a sequence of observations, the goal is often to find the most likely sequence of hidden states (the **decoding problem**) or to estimate the parameters of the HMM (the **training problem**).

Applications of HMMs in Bioinformatics

- **Gene Prediction:** One of the most common applications of HMMs in bioinformatics is **gene prediction** in DNA sequences. In this case, the hidden states could correspond to different parts of a gene (e.g., **exons, introns, coding regions, non-coding regions**), while the observations could be the nucleotides in the sequence. The model can be trained on annotated sequences to learn the transition and emission probabilities, allowing it to predict where genes are located in unannotated sequences. **Example:** A gene is typically composed of exons (coding regions) and introns (non-coding regions). HMMs can be used to distinguish between exonic and intronic regions based on patterns in nucleotide sequences.
- **Sequence Alignment and Profile Hidden Markov Models (pHMMs):** HMMs are also used in sequence alignment and in constructing profile HMMs. A profile HMM is a type of HMM that is trained on a multiple sequence alignment and used to detect homologous sequences. It can capture the typical patterns of amino acids or nucleotides in conserved regions across related sequences. **Example:** Profile HMMs are commonly used in tools like

HMMER for searching sequence databases. They are especially useful in detecting conserved domains and motifs across large families of sequences, such as in protein family classification.

- **RNA Secondary Structure Prediction**: HMMs are also used in RNA secondary structure prediction, where the hidden states represent different structural states (e.g., stem, loop, etc.), and the observations are the nucleotides in the RNA sequence. The model helps predict the most likely secondary structure of an RNA molecule based on observed sequence data.
- **Protein Structure Prediction**: HMMs are employed to predict protein structures, where the states may correspond to structural elements like alpha-helices, beta-sheets, or loops. By analyzing patterns in protein sequences and their known structures, HMMs can help predict the structural conformation of new proteins.
- **Motif Discovery**: In motif discovery, HMMs are used to model conserved sequence motifs within DNA, RNA, or protein sequences. These motifs may represent biologically significant patterns such as transcription factor binding sites or functional domains within proteins.

Algorithms for HMMs

There are three fundamental problems that need to be solved when working with HMMs: the **evaluation problem**, the **decoding problem**, and the **learning problem**. The main algorithms used to address these problems are:

- **Forward Algorithm (Evaluation)**: The forward algorithm computes the probability of observing a given sequence of observations given the model parameters. It uses dynamic programming to efficiently sum over all possible state sequences.
- **Viterbi Algorithm (Decoding)**: The Viterbi algorithm finds the most likely sequence of hidden states given an observation sequence. It is also based on dynamic programming and ensures that the path with the highest probability is selected.
- **Baum-Welch Algorithm (Learning/Training)**: The Baum-Welch algorithm is used to train an HMM, i.e., to estimate the transition and emission probabilities from a set of observed sequences. It is a special case of the **Expectation-Maximization (EM)** algorithm.

Advantages and Limitations of HMMs

Advantages:

- **Flexibility**: HMMs can model a wide variety of sequential data types, including nucleotide sequences, protein sequences, and structural data.
- **Powerful Tool**: They are capable of modeling hidden dependencies and probabilistic transitions between different states, making them particularly useful for complex biological systems.
- **Effective in Sequence Analysis**: HMMs have been proven to be extremely effective in tasks like gene prediction, sequence alignment, and motif detection, often providing state-of-the-art performance.

Limitations:

- **Assumption of Markov Property**: HMMs assume that the system's state depends only on the previous state (first-order Markov assumption). This may not always hold in biological systems, where longer-range dependencies (e.g., between distant nucleotides or residues) can be important.
- **Computational Complexity**: HMMs can be computationally expensive to train and decode, especially with large datasets.
- **Need for Sufficient Data**: HMMs require a large amount of annotated data to accurately learn the model parameters, which can be a limitation when training on rare or poorly understood biological systems.

MACHINE LEARNING APPROACHES

Machine learning (ML) is a branch of artificial intelligence that focuses on building algorithms that allow computers to learn from and make predictions or decisions based on data. In bioinformatics, machine learning techniques are used to extract valuable insights from complex and large biological datasets. These approaches have become essential tools for a wide range of applications, from sequence analysis and protein structure prediction to drug discovery and personalized medicine. Machine learning in bioinformatics involves using computational methods to find patterns, classify biological data, predict outcomes, or cluster data points based on shared characteristics. Given the vast amounts of data generated in genomics, proteomics, and other biological fields, ML techniques help in making sense of this data, automating tasks, and improving predictions.

TYPES OF MACHINE LEARNING APPROACHES

Machine learning can be broadly categorized into three types based on how the model is trained: supervised learning, unsupervised learning, and reinforcement learning. There are also other techniques, such as semi-supervised learning and deep learning, which have gained significant attention in bioinformatics.

Supervised Learning

In supervised learning, a model is trained on labeled data, where the input data points are paired with the correct output (the target or label). The goal of the model is to learn the mapping between inputs and outputs, which can then be used to predict labels for new, unseen data.

Applications in Bioinformatics:

- **Gene Expression Prediction:** ML models can be trained to predict gene expression levels based on microarray or RNA-Seq data. Features such as nucleotide sequences, gene annotations, and regulatory elements can be used to predict gene activity in different conditions.
- **Classification of Disease States:** For example, models can classify patients as having different types of cancer based on their genetic or molecular profiles. A classifier (such as support vector machines or random forests) is trained on a dataset with known disease categories and learns to predict the disease status for new patients.
- **Protein Function Prediction:** Supervised learning can be used to classify protein sequences based on their functions. Data on protein sequences, structures, and known functional annotations can be used to build models that predict the biological functions of newly discovered proteins.

Common Algorithms:

- **Support Vector Machines (SVM):** These are widely used for classification tasks, such as predicting disease types or classifying gene sequences.
- **Random Forests:** A robust, ensemble learning method often used for classification and regression tasks, such as predicting gene expression or disease outcomes.
- **Neural Networks:** Used for complex, non-linear relationships, especially when dealing with large datasets like images or high-dimensional biological data.
- **K-Nearest Neighbors (KNN):** Used for classification problems, such as classifying protein structures based on sequence similarity.

Unsupervised Learning

Unsupervised learning involves training a model on data that has no labeled output. The goal is to find hidden patterns or structures in the data without any predefined categories or labels. This is often used to explore biological data and group similar samples or sequences.

Applications in Bioinformatics:

- **Clustering of Gene Expression Data**: Clustering algorithms (such as k-means or hierarchical clustering) are used to group genes with similar expression profiles across different conditions. This can help identify genes that are co-expressed and might function together in biological pathways.
- **Protein Structure Prediction**: Unsupervised learning can be used to group proteins with similar structural features or predict structural motifs from sequence data.
- **Discovery of Hidden Patterns**: In genomics, unsupervised methods can help identify patterns or motifs in DNA sequences that are not well understood, such as conserved regulatory regions or novel biomarker sequences.

Common Algorithms:

- **K-Means Clustering**: A widely used method to partition data into k clusters, where each data point belongs to the cluster with the nearest mean. It is often used in gene expression analysis and to identify patterns in large omics datasets.
- **Hierarchical Clustering**: Creates a dendrogram that represents a nested series of clusters. It's commonly used in gene expression analysis and when exploring protein families or gene families.
- **Principal Component Analysis (PCA)**: Although often used as a dimensionality reduction technique, PCA can also be used for unsupervised learning to explore high-dimensional data, such as genetic sequences or protein structures, by reducing the number of features while preserving most of the variance in the data.

Reinforcement Learning

Reinforcement learning (RL) is an area of machine learning where agents learn by interacting with an environment and receiving feedback in the form of rewards or penalties. RL models are used to optimize decision-making strategies by learning from the consequences of actions taken.

Applications in Bioinformatics:

- **Drug Discovery**: RL can be used to optimize drug design by training models to select chemical compounds that are most likely to bind to a target protein. By simulating the interactions between molecules and their targets, RL algorithms can suggest novel molecules that are more likely to have therapeutic effects.

- **Protein Folding:** RL approaches have been explored for predicting protein folding, where the model learns to predict the most stable structure of a protein given its amino acid sequence. This is one of the more complex applications of RL in bioinformatics.

Example Algorithms:

- **Q-Learning:** An algorithm used in RL that helps the agent learn the value of actions in a given state, optimizing the long-term reward. This can be applied to optimize protein structure prediction or drug discovery.
- **Deep Q-Networks (DQNs):** These are deep learning models combined with reinforcement learning to learn strategies in high-dimensional spaces. They have been used for protein design and other complex bioinformatics t

Semi-supervised Learning

Semi-supervised learning lies between supervised and unsupervised learning. It uses a small amount of labeled data and a large amount of unlabeled data. This approach is especially useful in bioinformatics, where labeled data can be expensive or difficult to obtain.

Applications in Bioinformatics:

- **Cancer Subtyping:** In cancer genomics, semi-supervised learning can be used to identify subtypes of cancer from a small number of annotated samples (e.g., tumor vs. normal) and a larger set of unlabeled samples.
- **Gene Regulatory Networks:** Semi-supervised methods can help build gene regulatory networks by learning from partially labeled data, such as known transcription factors and target genes, along with large amounts of unlabeled gene expression data.

Common Algorithms:

- **Self-training:** A semi-supervised learning technique where a model is first trained on the labeled data, then used to predict labels for the unlabeled data. The most confident predictions are added to the training set.
- **Co-training:** A method where two models are trained on different feature subsets and can teach each other by labeling unlabeled data.

Deep Learning

Deep learning is a subset of machine learning that uses neural networks with many layers (hence the term "deep"). Deep learning models, particularly convolutional neural networks (CNNs) and recurrent neural networks (RNNs), have shown significant success in bioinformatics, particularly when dealing with large and high-dimensional biological data.

Applications in Bioinformatics:

- **Genomic Sequence Analysis:** CNNs and RNNs can be used to predict the function of genes, identify regulatory regions, and classify variants from genomic sequences (e.g., predicting disease-associated mutations from DNA sequences).
- **Protein Structure Prediction:** Deep learning models like AlphaFold, which use deep convolutional neural networks, have revolutionized protein structure prediction by accurately predicting the 3D structure of proteins based on their amino acid sequences.
- **Single-Cell RNA-Seq Analysis:** Deep learning methods are used to analyze single-cell RNA sequencing data, helping to identify cell types and infer gene regulatory networks.

Common Algorithms:

- **Convolutional Neural Networks (CNNs):** These are particularly effective at analyzing grid-like data, such as images or sequence motifs in DNA, RNA, or protein sequences.
- **Recurrent Neural Networks (RNNs):** RNNs are useful for sequence data, such as time-series data or biological sequences, where the order of the data is important (e.g., DNA, RNA, protein sequences).
- **Autoencoders:** These are used for unsupervised learning tasks such as dimensionality reduction, anomaly detection, or generating new biological data, like synthetic gene sequences.

Challenges and Future Directions

While machine learning has made tremendous advances in bioinformatics, several challenges remain:

- **Data Quality and Quantity:** Many biological datasets are noisy, incomplete, or imbalanced. ML models often require large, well-labeled datasets to achieve high accuracy, and acquiring such datasets in biology can be expensive or technically challenging.
- **Interpretability:** Many ML models, especially deep learning methods, are often seen as "black boxes," making it difficult to interpret how they make decisions. This is particularly important in bioinformatics, where interpretability is crucial for scientific discovery and clinical applications.
- **Generalization:** A model trained on one dataset might not always generalize well to others, especially in cases where biological systems are highly diverse or evolve rapidly.

Despite these challenges, machine learning in bioinformatics holds immense promise. With advancements in algorithms, more powerful computing infrastructure, and more comprehensive biological data, ML approaches will continue to play a critical role in fields like personalized medicine, drug discovery, genomic analysis, and systems biology.

10 Chapter

PHYLOGENETICS

INTRODUCTION

Phylogenetics is the study of the evolutionary relationships among species or groups of organisms. By analyzing genetic, morphological, and molecular data, scientists can reconstruct the evolutionary history of life on Earth, illustrating how different species are related through common ancestry. This field plays a crucial role in understanding the diversity of life, the mechanisms of evolution, and the patterns of speciation over time. Phylogenetic trees, or cladograms, are commonly used to visually represent these relationships, helping researchers to trace the origins and divergence of species across different time scales.

Phylogenetic Methods in Bioinformatics

Sequence Alignment: The first step in many phylogenetic analyses is to align sequences from different species or strains. This is crucial because the degree of similarity between sequences can provide insights into their evolutionary relationships. Multiple sequence alignment (MSA) algorithms like ClustalW, MUSCLE, and MAFFT are widely used for this task. These algorithms align nucleotide or protein sequences to identify homologous regions, which are then used to build phylogenetic trees.

Distance-based Methods: These methods calculate pairwise distances between sequences based on the number of differences between them. The most common distance measures include *p-distance* (proportion of differing nucleotides) or more sophisticated models like *Jukes-Cantor* or *Kimura 2-parameter* that account for different rates of substitutions across sites. The calculated distances are used to generate phylogenetic trees using algorithms such as Neighbor-Joining (NJ) or Unweighted Pair Group Method with Arithmetic Mean (UPGMA).

Character-based Methods (Maximum Parsimony and Maximum Likelihood)

- **Maximum Parsimony (MP):** This method seeks the tree that minimizes the total number of evolutionary changes (substitutions) required to explain the observed sequence data. It is a more direct approach based on character changes rather than sequence distances. Tools like PAUP* and TNT are used for MP analysis.

- **Maximum Likelihood (ML):** Maximum likelihood methods, such as those implemented in RAxML and PhyML, calculate the probability of the data given a tree, and then find the tree that maximizes this likelihood. These methods are considered more statistically rigorous and can accommodate complex evolutionary models (e.g., nucleotide substitution models), making them suitable for large datasets.

Bayesian Inference: Bayesian methods are another approach for constructing phylogenetic trees. Unlike ML, which seeks to maximize the likelihood of the data, Bayesian methods calculate the probability distribution of trees based on prior assumptions about evolutionary processes. Tools like Mr Bayes and BEAST use Markov Chain Monte Carlo (MCMC) methods to sample from the posterior distribution of trees. Bayesian methods can incorporate uncertainty about tree topology and model parameters, making them particularly useful for more complex or uncertain datasets.

Phylogenetic Networks: While traditional phylogenetic trees represent a single evolutionary pathway, phylogenetic networks (such as split networks or reticulated networks) are used to represent more complex evolutionary histories, especially when gene flow or horizontal gene transfer (HGT) occurs. These networks allow for the representation of conflicting phylogenetic signals, which might occur in cases of incomplete lineage sorting or hybridization.

Tools and Software for Phylogenetic Analysis in Bioinformatics

Several bioinformatics tools and software packages are used for conducting phylogenetic analyses:

- **MEGA (Molecular Evolutionary Genetics Analysis):** A user-friendly tool for performing sequence alignment, distance-based phylogeny, maximum likelihood, and bootstrap analysis.
- **RAxML (Randomized Axelerated Maximum Likelihood):** A powerful tool for large-scale maximum likelihood phylogenetic inference, commonly used in genome-scale analyses.
- **BEAST (Bayesian Evolutionary Analysis Sampling Trees):** A tool for Bayesian analysis, often used for molecular dating and phylogenetic inference with temporal data.
- **MrBayes:** A software package for Bayesian inference of phylogeny, widely used for constructing posterior distributions of trees using MCMC.
- **Clustal Omega:** A widely used sequence alignment tool, often employed in combination with phylogenetic software to prepare data for tree construction.

- **PhyML:** A software for maximum likelihood phylogenetic analysis, capable of handling large datasets and various evolutionary models.

Applications of Phylogenetics in Bioinformatics

- **Evolutionary Biology and Comparative Genomics**: Phylogenetic trees allow researchers to trace the evolutionary history of species or genes, identifying conserved genetic regions and studying how genetic variation contributes to phenotypic diversity. For example, phylogenetic approaches are crucial in tracing the evolution of drug-resistant strains in pathogens, such as *Mycobacterium tuberculosis*.
- **Taxonomy and Species Identification**: Phylogenetic analysis can help clarify the relationships among species, guiding classification schemes and resolving taxonomic ambiguities. By comparing genetic sequences, scientists can identify new species or clarify the phylogenetic placement of poorly understood groups.
- **Functional Genomics and Gene Evolution**: Phylogenetic analysis can be used to study gene family evolution, tracing the origins and functional divergence of genes across species. For example, researchers can analyze gene trees to infer the evolutionary pressures on specific gene families or regulatory regions.
- **Metagenomics and Microbial Ecology**: In microbial ecology, phylogenetic methods are applied to metagenomic datasets to understand the diversity and evolutionary history of microbial communities. Phylogenetic analysis of 16S rRNA genes or other markers allows for the identification and classification of microorganisms in complex environmental samples.
- **Viral Evolution and Epidemiology**: Phylogenetic techniques are essential for tracking the evolution of viruses, such as in the case of influenza, HIV, and SARS-CoV-2. By constructing viral phylogenies, researchers can understand transmission dynamics, mutation rates, and identify viral strains responsible for outbreaks.

Challenges and Future Directions

Despite its powerful applications, phylogenetic analysis in bioinformatics faces several challenges:

- **Data Quality and Completeness:** The quality and completeness of sequence data can affect the accuracy of phylogenetic inferences. Errors in sequencing, incomplete genomes, or unrepresented lineages can distort phylogenetic trees.

- **Computational Complexity:** Large genomic datasets, especially those involving thousands of species or genes, require significant computational resources for analysis. Phylogenetic tree construction can become computationally intensive, particularly for maximum likelihood or Bayesian methods.
- **Model Selection:** Choosing the appropriate evolutionary model is crucial for accurate phylogenetic reconstruction. Poor model selection can lead to incorrect conclusions about evolutionary relationships.
- **Horizontal Gene Transfer (HGT):** In cases of HGT, where genes are transferred between unrelated species, traditional phylogenetic trees may be misleading, necessitating the use of more sophisticated methods like phylogenetic networks.

Moving forward, improvements in sequencing technologies, increased computational power, and the development of more advanced algorithms will continue to enhance the resolution and accuracy of phylogenetic analyses. Additionally, incorporating multi-omics data (e.g., transcriptomics, proteomics) and refining methods to account for complex evolutionary processes like gene flow, coevolution, and hybridization will broaden the scope of phylogenetics in bioinformatics.

PHYLOGENETIC TREES AND EVOLUTION

A **phylogenetic tree**, also known as an **evolutionary tree** or **tree of life**, is a branching diagram that represents the evolutionary relationships among a set of organisms or genes. The tree illustrates how different species or genetic sequences have diverged from common ancestors over time, offering a visual representation of their evolutionary history. By analyzing these trees, scientists can infer the patterns of descent, divergence, and adaptation that have shaped the diversity of life on Earth.

Structure of a Phylogenetic Tree

A phylogenetic tree consists of several key components:

1. **Root**: The base of the tree represents the common ancestor of all the species or sequences in the tree. The root indicates the most recent common ancestor (MRCA) of the group being studied.
2. **Nodes**: These are the branching points on the tree, representing the most recent common ancestors (MRCA) of the groups descending from them. Each node represents a speciation event or divergence.
3. **Branches**: The lines connecting nodes represent the evolutionary paths that lead to different species or genes. The length of the branches may represent

genetic divergence or time, depending on the tree-building method used.

4. **Tips (or Leaves):** These are the endpoints of the tree, representing the current species or genes being studied. The tips are the most recent descendants of the ancestral nodes.
5. **Clades:** A clade is a group of organisms that includes a common ancestor and all its descendants. Clades are the main units of phylogenetic trees and can vary in size from a single species to entire groups of organisms (e.g., all mammals).

Types of Phylogenetic Trees

- **Cladogram:** A simple tree where the branch lengths have no significance, and only the branching pattern (the order of divergence) is important. It shows the relative relationships between taxa.
- **Phylogram:** A tree where the branch lengths are proportional to the amount of evolutionary change (e.g., genetic distance) between nodes.
- **Chronogram:** A tree where the branch lengths are scaled to represent time, showing the temporal relationships between different species based on molecular clock data.

Common Ancestry and Divergence

Phylogenetic trees reflect the principle of common descent—the idea that species share a common ancestor. The closer two species are in a phylogenetic tree, the more recent their common ancestor, and the more closely related they are in an evolutionary sense. For example, humans and chimpanzees share a relatively recent common ancestor, making them more closely related to each other than either is to, say, a mouse or a frog. Divergence occurs when a population of organisms is split by geographical, environmental, or other barriers, leading to genetic isolation and eventual speciation. As a result, different evolutionary paths are followed by each branch of the tree.

Evolutionary Relationships and Traits

By studying the branching patterns of a phylogenetic tree, scientists can infer the evolutionary relationships between species and trace the evolution of specific traits, such as the development of wings in birds, the loss of flight in certain species, or the evolution of antibiotic resistance in bacteria. The shared traits or genetic similarities between species reflect their evolutionary history and common ancestry.For instance, in a tree of mammals, the presence of mammary glands is a shared feature for all mammals, and analyzing the tree can show how this trait evolved in the common ancestor of mammals.

Molecular Evolution and Sequence Comparison

In modern phylogenetics, molecular data, such as DNA, RNA, or protein sequences, are used to build more accurate trees. Evolution is reflected in the genetic changes that accumulate over time within populations, and these genetic differences are key to understanding phylogenetic relationships. For example, the comparison of specific genes (such as rRNA genes) across different species can reveal their evolutionary relatedness. Over time, mutations in these genes accumulate in a way that allows scientists to estimate the time of divergence between species, helping to construct a molecular **clock** that can be used to date evolutionary events.

Macroevolution and Microevolution

- **Macroevolution** refers to large-scale evolutionary changes, such as the emergence of new species, the formation of new genera or families, or the transition from one major group of organisms to another (e.g., from reptiles to birds). Phylogenetic trees capture these large-scale events, showing how major groups of organisms are related.
- **Microevolution** involves smaller-scale changes within species, such as changes in allele frequencies in a population over time. While phylogenetic trees typically focus on broader evolutionary patterns, they can also be used to understand how smaller evolutionary changes accumulate and lead to speciation.

Horizontal Gene Transfer and Evolutionary Complexity

In addition to vertical evolution (the inheritance of genetic material from parent to offspring), horizontal gene transfer (HGT) is an important evolutionary mechanism, particularly in microorganisms like bacteria. HGT occurs when genes are transferred between unrelated species, complicating the tree-like structure of evolution. In these cases, phylogenetic trees may need to be supplemented with phylogenetic networks to account for reticulate evolution and shared genetic material that does not follow the usual parent-to-offspring inheritance model.

Applications of Phylogenetic Trees in Evolutionary Studies

- **Tracing the Evolution of Species:** Phylogenetic trees provide insights into the evolutionary history of species by reconstructing their ancestry and relationships. For example, by examining the phylogenetic relationships between different species of mammals, scientists can infer the common ancestor of modern whales, dolphins, and land-dwelling mammals like cows and deer.

- **Understanding Evolution of Specific Traits:** Phylogenetic trees help track the evolution of specific biological traits. For example, in studying the evolution of flight in birds, the tree can show how flight evolved in a common ancestor of birds, while other lineages, like flightless ostriches or penguins, represent distinct evolutionary paths that lost this ability.
- **Conservation Biology and Phylogeny:** Phylogenetic analysis is also used in conservation biology to identify evolutionary distinct species that may need special attention. By preserving species that are evolutionarily unique, conservation efforts can help protect genetic diversity that may be crucial for future adaptation.
- **Epidemiology and Pathogen Evolution:** Phylogenetic trees are extensively used in epidemiology to track the evolution and spread of pathogens. For instance, during the COVID-19 pandemic, phylogenetic trees were used to trace the origins and global transmission patterns of the SARS-CoV-2 virus, helping public health officials track outbreaks and predict trends.

Molecular Clocks and Dating Evolutionary Events: Molecular clocks use the rate of genetic mutations to estimate the timing of evolutionary events. By calibrating the clock with known fossil data or geological events, scientists can date the divergence of species, providing insight into the timing of key evolutionary transitions (e.g., the split between humans and chimpanzees, or the origin of flowering plants).

Phylogenetic trees are a foundational tool in evolutionary biology, offering a means to visualize the relationships between species or genes and to study how life has evolved over time. They represent the concept of common ancestry and descent with modification—the key principles behind the theory of evolution. By integrating molecular, morphological, and genetic data, phylogenetic trees help scientists answer important questions about the origins of life, the diversification of species, and the mechanisms driving evolutionary change. Understanding phylogenetic trees not only enhances our knowledge of life's history but also provides a framework for exploring future research in fields such as genomics, epidemiology, conservation, and comparative biology.

PHYLOGENETIC INFERENCE METHODS

Phylogenetic inference is the process of deducing the evolutionary relationships among a group of species or genes based on molecular or morphological data. The goal is to reconstruct the evolutionary tree (phylogenetic tree) that best represents the common ancestry and divergence of the taxa in question. Various methods have been developed to infer phylogenies, each with its strengths and limitations. These methods can be broadly categorized into distance-based methods, character-based methods, and probabilistic methods. Here is a detailed description of the most commonly used phylogenetic inference methods:

Distance-Based Methods

Distance-based methods calculate the evolutionary distance between pairs of sequences (species or genes) based on their sequence differences. These methods are simpler and faster than other approaches, but they make assumptions about how changes accumulate over time and may not capture complex evolutionary processes well.

Methods:

Neighbor-Joining (NJ): Neighbor-Joining is one of the most popular distance-based methods for constructing phylogenetic trees. It starts by constructing a matrix of pairwise distances between the taxa being studied. The algorithm then iteratively joins pairs of taxa with the smallest pairwise distance, forming nodes, and recalculates distances between newly formed groups. This process continues until all taxa are grouped into a tree.

Advantages:

- Simple and computationally efficient.
- Suitable for large datasets.

Limitations:

- Assumes a constant rate of evolution, which may not be accurate for all datasets.
- Can be inaccurate if evolutionary rates vary significantly across the taxa being studied.

Unweighted Pair Group Method with Arithmetic Mean (UPGMA)

- UPGMA is another distance-based method that uses a hierarchical clustering approach. The method assumes that the rate of evolution is constant over time (a molecular clock) and uses pairwise distances to merge taxa into progressively larger clusters, ultimately resulting in a tree.

Advantages:

- Simple and intuitive.
- Efficient for smaller datasets with relatively constant evolutionary rates.

Limitations:

- Like NJ, it assumes a constant evolutionary rate, which may lead to inaccuracies if this assumption is violated.

- Less effective for datasets with unequal evolutionary rates across taxa.

Character-Based Methods

Character-based methods build phylogenetic trees by examining specific characters (traits or sequences) that vary among taxa. These methods consider the evolutionary changes in these characters as evidence of the relationships among species.

Methods:

Maximum Parsimony (MP): Maximum Parsimony is a method that seeks the tree with the fewest evolutionary changes (i.e., the minimum number of character state changes) required to explain the observed data. This method examines all possible tree topologies and selects the one that minimizes the total number of changes (substitutions, insertions, or deletions) across the entire tree.

Advantages:

- Simple and intuitive.
- Well-suited for datasets with relatively few changes and fewer taxa.

Limitations:

- MP may be prone to overfitting, particularly when there are large amounts of homoplasy (similar traits due to convergent evolution).
- The method does not take into account the probabilities of different evolutionary events, so it can be inaccurate for complex datasets.

Maximum Likelihood (ML): Maximum Likelihood is a statistical method that calculates the probability of the observed data given a particular phylogenetic tree. It does this by considering the possible evolutionary model (e.g., nucleotide substitution model) and the tree's topology. The tree with the highest likelihood (the best fit to the data) is selected as the inferred phylogeny. ML methods can accommodate a wide variety of models for molecular evolution and account for unequal rates of evolution across sites.

Advantages:

- Can incorporate complex models of evolution (e.g., rate variation across sites, different models for nucleotides or amino acids).
- More statistically rigorous than simpler methods like MP.
- Suitable for large datasets and detailed evolutionary questions.

Limitations:

- Computationally expensive, especially for large datasets.
- Requires a well-chosen evolutionary model to ensure accurate results.

Bayesian Inference (BI): Bayesian methods are based on probability theory and involve calculating the posterior probability distribution of phylogenetic trees given the data and prior assumptions about evolutionary processes. In Bayesian phylogenetic inference, the likelihood of a tree is combined with prior distributions (i.e., assumptions about tree topologies, substitution rates, etc.) to estimate the probability of different phylogenies.

Advantages:

- Provides a full probability distribution of trees, allowing for the incorporation of uncertainty in the analysis.
- Allows for the inclusion of prior knowledge (e.g., constraints on tree shape, models of evolution).
- Can handle complex datasets and multiple types of data (e.g., sequence data, morphological traits).

Limitations:

- Computationally intensive, particularly when using complex models and large datasets.
- Results can be sensitive to prior assumptions and model choice.

Probabilistic Methods

Probabilistic methods use statistical models to estimate the most likely tree based on observed data, and they usually allow for the incorporation of more complex evolutionary processes.

Methods:

Markov Chain Monte Carlo (MCMC) Methods: MCMC is a class of algorithms that generates a sample of trees (or tree topologies) from the posterior distribution using random sampling. In phylogenetic inference, this is typically used in Bayesian methods, where the tree is iteratively refined based on the likelihood of the data and the prior distribution.

Advantages:

- MCMC methods provide a robust way to estimate the uncertainty in tree topologies.
- Can incorporate various evolutionary models and complex datasets.

Limitations:

- Computationally expensive, requiring significant processing power and time.
- Results can be influenced by the choice of priors and the convergence of the Markov chain.

Coalescent Theory: Coalescent-based methods model the genealogy of a sample of sequences by tracing the most recent common ancestor of sampled alleles. These methods are particularly useful in molecular evolution studies where the goal is to infer evolutionary processes like genetic drift, population structure, and species divergence over time.

Advantages:

- Useful for estimating divergence times and population dynamics.
- Accounts for population-level processes like bottlenecks and expansions.

Limitations:

- Requires large amounts of data and sophisticated models to achieve accurate results.
- Can be sensitive to assumptions about population size and structure.

Hybrid Methods

Hybrid methods combine the strengths of different phylogenetic inference approaches. For example, some methods use a distance matrix derived from sequence data but then apply maximum likelihood to refine the tree's topology. Hybrid approaches are increasingly common as they can offer more flexibility and accuracy.

- **Example:** Maximum Likelihood (ML) with Neighbor-Joining (NJ) or Bayesian Inference (BI) with Maximum Parsimony (MP). These hybrid methods can be used to balance the strengths of different approaches (e.g., computational efficiency with NJ or UPGMA and statistical rigor with ML).

Phylogenetic Networks

In certain cases, the evolutionary relationships between taxa cannot be accurately represented by a tree due to complex evolutionary processes such as horizontal gene transfer or reticulate evolution (where two species hybridize and form a new lineage). In such cases, phylogenetic networks can be used to represent the non-tree-like relationships among organisms.

Methods for Network Construction:

- NeighborNet: A method for constructing phylogenetic networks using distance-based data.
- Split Networks: These are graphical representations that show the degree of conflict or uncertainty in the tree-building process.

Advantages:

- Can capture complex evolutionary relationships that are not well represented by simple trees.

Limitations:

- More complex than tree-based methods and require specialized algorithms and software.
- Can be computationally expensive and challenging to interpret.

Phylogenetic inference methods are essential tools for reconstructing the evolutionary relationships between species, genes, and other taxa. Each method has its strengths, from the simplicity and speed of distance-based methods like Neighbor-Joining, to the statistical rigor and model flexibility of Maximum Likelihood and Bayesian methods. The choice of method depends on the dataset, the research question, and the computational resources available. By combining different methods and refining models over time, researchers continue to improve the accuracy and efficiency of phylogenetic analyses, enhancing our understanding of evolutionary processes.

DISTANCE-BASED, CHARACTER-BASED, AND BAYESIAN METHODS

Phylogenetic inference methods are crucial tools in evolutionary biology, used to reconstruct the evolutionary relationships between species or genes. These methods can be classified into different categories based on how they approach the analysis of evolutionary data. The three major categories are distance-based methods, character-based methods, and Bayesian methods. Each of these methods has its own strengths, assumptions, and applications. Here's a detailed overview of these methods.

Distance-Based Methods

Distance-based methods focus on calculating the evolutionary distance between pairs of taxa (species, genes, etc.) based on their genetic or morphological differences. These methods are relatively simple and computationally efficient, making them useful for large datasets, though they have limitations in terms of their assumptions about evolutionary processes.

- **Neighbor-Joining (NJ) Method:**The Neighbor-Joining method is a distance-based approach that constructs a phylogenetic tree by iteratively joining pairs of taxa that have the smallest pairwise distance. It starts by calculating a matrix of pairwise distances (e.g., genetic distances) between all taxa. In each iteration, the two taxa (or groups of taxa) with the smallest distance are merged to form a new node, and the distance matrix is updated. The process continues until all taxa are grouped into a single tree.
- **Assumptions:** NJ assumes that the distance matrix reflects the true evolutionary distances between taxa, but it does not require a model of evolution. It works well when evolutionary rates are roughly constant across all taxa.

Advantages:

- Fast and computationally efficient, making it suitable for large datasets.
- No need for complex models of molecular evolution, making it easier to implement.

Limitations:

- Assumes constant rates of evolution across lineages (a molecular clock), which is not always true.
- Does not take into account the different probabilities of various types of substitutions, which can lead to inaccuracies in certain cases.
- The resulting tree may not always be the most accurate if the assumptions are violated.

Unweighted Pair Group Method with Arithmetic Mean (UPGMA)

UPGMA is another distance-based clustering method that builds phylogenetic trees by progressively joining pairs of taxa based on their pairwise distances. Like NJ, it uses a distance matrix, but it assumes a molecular clock (i.e., constant rates of evolution across all lineages).

- **Assumptions:** UPGMA assumes that the evolutionary rate is the same for all taxa (molecular clock hypothesis), which can lead to errors when this assumption is violated.

Advantages:

- Simple and easy to compute.
- Suitable for small datasets or when the molecular clock assumption is valid.

Limitations:

- Assumes constant evolutionary rates across all taxa, which is rarely true in nature.

- Less accurate than more sophisticated methods like Maximum Likelihood (ML) or Bayesian Inference.

Character-Based Methods

Character-based methods are more sophisticated than distance-based methods, as they consider the actual sequence or trait data of individual characters (e.g., nucleotide positions in DNA, amino acid positions in proteins). These methods evaluate the best phylogeny by comparing how many changes (mutations) in characters are needed to explain the observed data.

Character-Based Methods:

Maximum Parsimony (MP): Maximum Parsimony seeks to find the tree topology that minimizes the total number of evolutionary changes (character state changes) across all characters. In MP, each possible tree is evaluated, and the tree with the least number of changes (substitutions, insertions, or deletions) is selected as the best tree.

- **Assumptions:** MP assumes that the simplest explanation (fewest changes) is the most likely. This is known as the parsimony principle.

Advantages:

- Intuitive and relatively simple to apply.
- Can work with both molecular and morphological data.
- Good for small datasets where changes are infrequent.

Limitations:

- Homoplasy: The occurrence of similar traits due to convergent evolution can mislead the tree-building process (i.e., traits that evolve independently in different lineages may lead to incorrect conclusions).
- Does not incorporate the rates of evolutionary changes or the probability of different types of changes.
- Computationally expensive for large datasets, as the number of possible tree topologies increases exponentially with more taxa.

Maximum Likelihood (ML): Maximum Likelihood methods calculate the probability of observing the data (e.g., sequence alignments) given a specific tree topology and a chosen model of molecular evolution. The goal is to find the tree that maximizes the likelihood of the observed data under the given model.

Assumptions: ML requires a predefined model of evolution (e.g., a nucleotide substitution model such as Jukes-Cantor, Kimura, etc.). It assumes that the data follow this model of evolution.

Advantages:

- Incorporates a model of evolution, which allows it to account for varying rates of substitution across sites.
- More statistically rigorous than MP, providing better estimates of phylogenetic relationships.
- Can handle large datasets with more taxa and more complex models.

Limitations:

- Computationally intensive, especially with large datasets or complex models.
- Requires careful model selection. Incorrect models can lead to inaccurate trees.
- Does not inherently estimate statistical support for individual nodes (although bootstrapping or other techniques can be used to provide support).

Bayesian Methods

Bayesian methods use a probabilistic framework to estimate phylogenies, combining the observed data with prior information about the evolutionary process. Unlike other methods, Bayesian methods generate a distribution of trees (a posterior distribution) and provide a probability for each possible tree, allowing researchers to assess the uncertainty in the phylogenetic tree.

Features of Bayesian Methods:

Bayesian Inference (BI): In Bayesian Inference, a phylogenetic tree is generated by calculating the probability of a tree (posterior probability) given the data (likelihood) and prior knowledge (prior probability). The method uses Markov Chain Monte Carlo (MCMC) simulations to sample from the posterior distribution of trees, iterating until the chains converge to a stable distribution.

- **Assumptions:** Bayesian methods assume a chosen model of evolution (e.g., a nucleotide substitution model) and combine this with prior assumptions (e.g., tree shape, branch length) to calculate the posterior probability of different trees.

Advantages:

- Provides a probability distribution of trees, offering insight into the uncertainty of tree topology.
- Can incorporate complex evolutionary models, such as rate variation across sites or among lineages.

- Can incorporate prior information, such as previous phylogenetic knowledge or constraints on tree topology (e.g., the tree must be rooted).
- Bayesian methods often provide posterior probabilities for the nodes of the tree, offering a measure of support for each branching point.

Limitations:

- Computationally intensive, particularly for large datasets or complex models.
- Requires appropriate prior information and model selection, which can influence results.
- MCMC requires a large number of iterations to achieve convergence and obtain reliable results.
- **Markov Chain Monte Carlo (MCMC):** MCMC is a key part of Bayesian inference. It is a method for sampling from a probability distribution by constructing a chain of possible tree topologies that converge to the true posterior distribution after many iterations.

Advantages:

- Provides a way to calculate confidence intervals (uncertainty) around tree estimates.
- Allows for the integration of complex models of molecular evolution.

Limitations:

- MCMC can be slow and computationally expensive, particularly when using complex models.
- The convergence of the MCMC chain needs to be assessed to ensure reliable results.

Comparison of the Methods

Method	Type	Strengths	Limitations
Distance-Based (NJ, UPGMA)	Distance-based	Fast, simple, works for large datasets	Assumes constant evolutionary rates, lacks statistical support
Character-Based (MP, ML)	Character-based	Accounts for actual changes in traits, more accurate	MP can suffer from homoplasy, ML is computationally intensive
Bayesian Methods	Probabilistic	Provides probability distributions of trees, handles complex models	Computationally expensive, requires prior knowledge and convergence checks

Each of the three phylogenetic inference methods—distance-based, character-based, and Bayesian—has unique advantages and limitations. Distance-based methods are simple and fast but may be inaccurate if evolutionary rates vary. Character-based methods (like Maximum Parsimony and Maximum Likelihood) offer more detailed insights into the evolutionary process, with ML providing more accurate and statistically rigorous results. Bayesian methods provide a robust framework for incorporating prior knowledge and estimating uncertainty in phylogenetic trees, making them ideal for complex datasets but computationally demanding. The choice of method depends on the dataset, research objectives, and computational resources available.

11 Chapter

PROTEIN STRUCTURE AND FUNCTION

Protein data bank

The Protein Data Bank (PDB) is a crystallographic database for the three- dimensional structural data of large biological molecules, such as proteins and nucleic acids. The data, typically obtained by X-ray crystallography, NMR spectroscopy, or, increasingly, cryo-electron microscopy, and submitted by biologists and biochemists from around the world, are freely accessible on the Internet via the websites of its member organisations (PDBe, PDBj, and RCSB). The PDB is overseen by an organization called the Worldwide Protein Data Bank, wwPDB.

The PDB is a key resource in areas of structural biology, such as structural genomics. Most major scientific journals, and some funding agencies, now require scientists to submit their structure data to the PDB. Many other databases use protein structures deposited in the PDB. For example, SCOP and CATH classify protein structures, while PDBsum provides a graphic overview of PDB entries using information from other sources, such as Gene ontology

The Protein Data Bank (PDB) at Brookhaven National Laboratory (BNL), is a database containing experimentally determined three-dimensional structures of proteins, nucleic acids and other biological macromolecules. The archives contain atomic coordinates, citations, primary and secondary structure information, crystallographic structure experimental data, as well as hyperlinks to many other scientific databases.

Protein Data Bank (PDB) format is a standard for files containing atomic coordinates. Structures deposited in the Protein Data Bank at the Research Collaboratory for Structural Bioinformatics (RCSB) are written in this standardized format. The complete PDB file specification provides for a wealth of information, including authors, literature references, and the identification of substructures such as disulfide bonds, helices, sheets, and active sites.

Protein Data Bank format consists of lines of information in a text file. Each line of information in the file is called a record. A file generally contains several different types of records, which are arranged in a specific order to describe a structure.

Selected Protein Data Bank Record Types

Record Type	
ATOM	atomic coordinate record containing the x,y,z orthogonal Angstrom coordinates for atoms in standard residues (amino acids and nucleic acids).
HETATM	atomic coordinate record containing the x,y,z orthogonal Angstrom coordinates for atoms in nonstandard residues. Nonstandard residues include inhibitors, cofactors, ions, and solvent. The only functional difference from ATOM records is that HETATM residues are by default not connected to other residues. Note that water residues should be in HETATM records.
TER	indicates the end of a chain of residues. For example, a hemoglobin molecule consists of four subunit chains which are not connected. TER indicates the end of a chain and prevents the display of a connection to the next chain.
SSBOND	defines disulfide bond linkages between cysteine residues.
HELIX	indicates the location and type (right-handed alpha, etc.) of helices. One record per helix.
SHEET	indicates the location, sense (anti-parallel, etc.) and registration with respect to the previous strand in the sheet (if any) of each strand in the model. One record per strand

PROTEIN ANALYSIS

The spatial arrangement of atoms in a protein is called a conformation. The term conformation refers to a structural state that can, without breaking any covalent bonds, interconvert with other structural states. A change in conformation could occur, for example, by rotation about single bonds. Of the innumerable conformations that are theoretically possible in a protein containing hundreds of single bonds, one generally predominates. This is usually the conformation that is thermodynamically the most stable, having the lowest Gibbs' free energy (G). Proteins in their functional conformation are called native proteins.

There are Four Levels of Architecture in Proteins

Conceptually, protein structure can be considered at four levels.

Primary structure includes all the covalent bonds between amino acids and is normally defined by the sequence of peptide-bonded amino acids and locations of disulfide bonds. The relative spatial arrangement of the linked amino acids is unspecified. Polypeptide chains are not free to take up any three-dimensional structure at random. Steric constraints and many weak interactions stipulate that some arrangements will be more stable than others.

Secondary structure refers to regular, recurring arrangements in space of adjacent amino acid residues in a polypeptide chain. There are a few common types of secondary structure, the most prominent being the a helix and the β conformation.

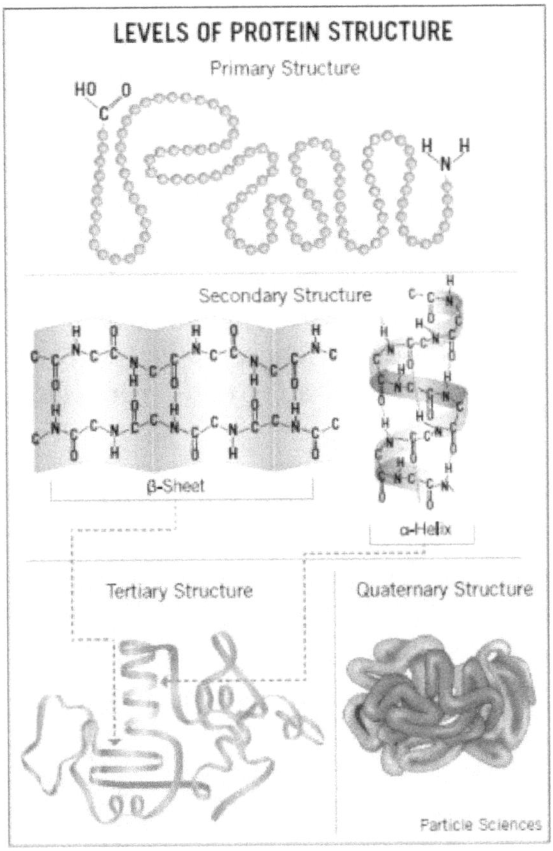

Levels of structure in proteins

Tertiary structure refers to the spatial relationship among all amino acids in a polypeptide; it is the complete three-dimensional structure of the polypeptide. The boundary between secondary and tertiary structure is not always clear. Several different types of secondary structure are often found within the three-dimensional structure of a large protein. Proteins with several polypeptide chains have one more level of structure: **quaternary structure**, which refers to the spatial relationship of the polypeptides, or subunits, within the protein.

Protein Secondary Structure

Several types of secondary structure are particularly stable and occur widely in proteins. The most prominent are the α helix and β conformations. Using fundamental chemical principles and a few experimental observations, Linus Pauling and Robert Corey predicted the existence of these secondary structures in 1951, several years before the first complete protein structure was elucidated.

In considering secondary structure, it is useful to classify proteins into two major groups: fibrous proteins, having polypeptide chains arranged in long strands or sheets, and globular proteins, with polypeptide chains folded into a spherical or globular shape. Fibrous proteins play important structural roles in the anatomy and physiology of vertebrates, providing external protection, support, shape, and form. They may constitute one-half or more of the total body protein in larger animals. Most enzymes and peptide hormones are globular proteins. Globular proteins tend to be structurally complex, often containing several types of secondary structure; fibrous proteins usually consist largely of a single type of secondary structure. Because of this structural simplicity, certain fibrous proteins played a key role in the development of the modern understanding of protein structure and provide particularly clear examples of the relationship between structure and function; they are considered in some detail after the general discussion of secondary structure.

THE PEPTIDE BOND IS RIGID AND PLANAR

In the peptide bond, the π-electrons from the carbonyl are delocalized between the oxygen and the nitrogen. This means that the peptide bond has ~40% double bond character. This partial double bond character is evident in the shortened bond length of the C–N bond. The length of a normal C–N single bond is 1.45 Å and a C=N double bond is 1.25 Å, while the peptide C–N bond length is 1.33 Å.

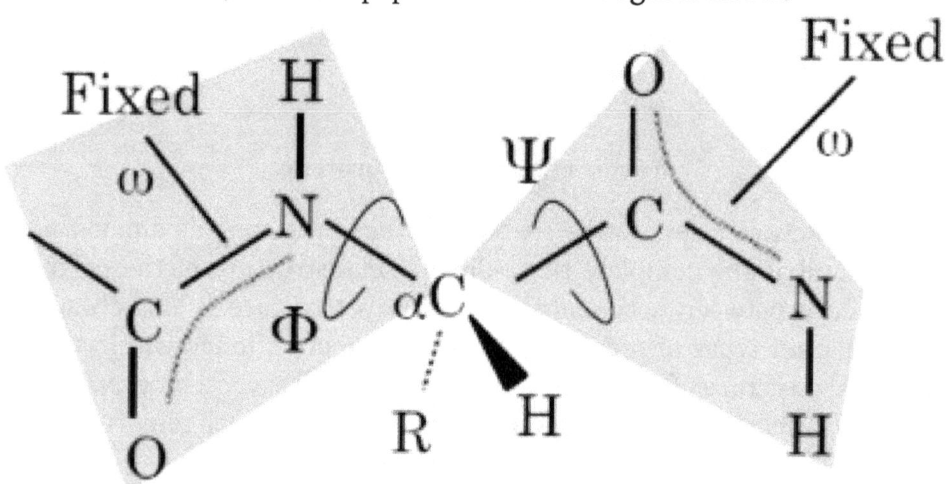

Peptide Bond – Dihedral angles

Because of its partial double bond character, rotation around the N–C bond is severely restricted. The peptide bond allows rotation about the bonds from the α- carbon, but not the amide C–N bond. Only the Φ and Ψ torsion angles can vary reasonably freely. In addition, the six atoms in the peptide bond (the two α-carbons, the amide O, and the amide N and H) are coplanar. Finally, the peptide bond has a dipole, with the O having a partial negative charge, and the N amide having a partial positive charge.

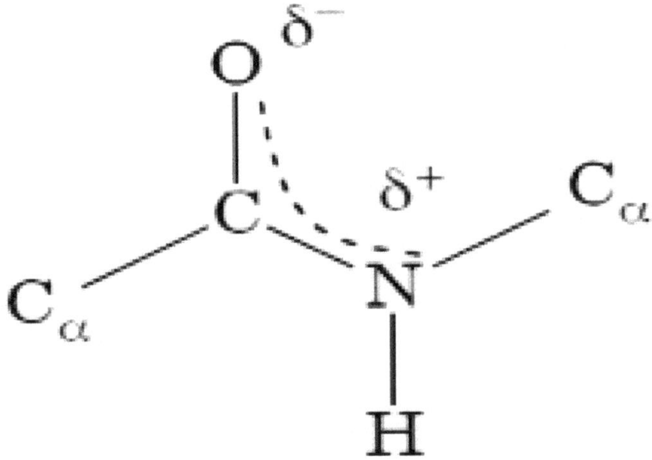

Peptide Bond – Dipole

This allows the peptide bond to participate in electrostatic interactions, and contributes to the hydrogen bond strength between the backbone carbonyl and the Namide proton.

Allowed values for Φ and Ψ are graphically revealed when Ψ is plotted versus Φ in a **Ramachandran plot**, introduced by G. N. Ramachandran.

The Ramachandran Plot

In a polypeptide the main chain N-Calpha and Calpha-C bonds relatively are free to rotate. These rotations are represented by the torsion angles phi and psi, respectively.

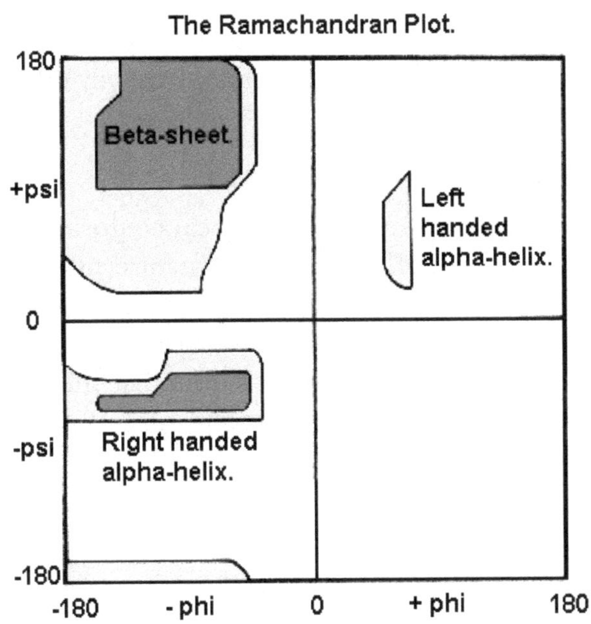

Ramachandran Plot

G N Ramachandran used computer models of small polypeptides to systematically vary phi and psi with the objective of finding stable conformations. For each conformation, the structure was examined for close contacts between atoms. Atoms were treated as hard spheres with dimensions corresponding to their van der Waals radii. Therefore, phi and psi angles which cause spheres to collide correspond to sterically disallowed conformations of the polypeptide backbone.

In the diagram above the white areas correspond to conformations where atoms in the polypeptide come closer than the sum of their van der Waals radii. These regions are sterically disallowed for all amino acids except glycine which is unique in that it lacks a side chain. The red regions correspond to conformations where there are no steric clashes, ie these are the allowed regions namely the alpha-helical and beta-sheet conformations. The yellow areas show the allowed regions if slightly shorter van der Waals radi are used in the calculation, ie the atoms are allowed to come a little closer together. This brings out an additional region which corresponds to the left-handed alpha-helix.

L-amino acids cannot form extended regions of left-handed helix but occasionally individual residues adopt this conformation. These residues are usually glycine but can also be asparagine or aspartate where the side chain forms a hydrogen bond with the main chain and therefore stabilizes this otherwise unfavorable conformation. The 3(10) helix occurs close to the upper right of the alpha-helical region and is on the edge of allowed region indicating lower stability.

Disallowed regions generally involve steric hindrance between the side chain C-beta methylene group and main chain atoms. Glycine has no side chain and therefore can adopt phi and psi angles in all four quadrants of the Ramachandran plot. Hence it frequently occurs in turn regions of proteins where any other residue would be sterically hindered.

Secondary structure

The term secondary structure refers to the local conformation of some part of a polypeptide. The discussion of secondary structure most usefully focuses on common regular folding patterns of the polypeptide backbone. A few types of secondary structure are particularly stable and occur widely in proteins. The most prominent are the α-helix and β-sheet. Using fundamental chemical principles and a few experimental observations, Pauling and Corey predicted the existence of these secondary structures in 1951, several years before the first complete protein structure was elucidated.

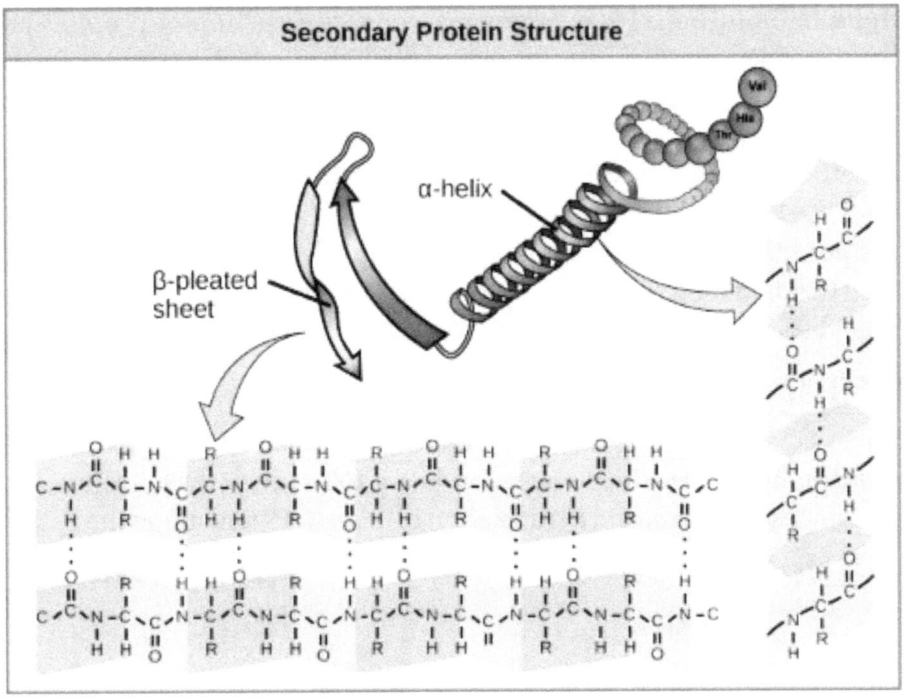

Secondary protein structures

Alpha helix (α-helix)

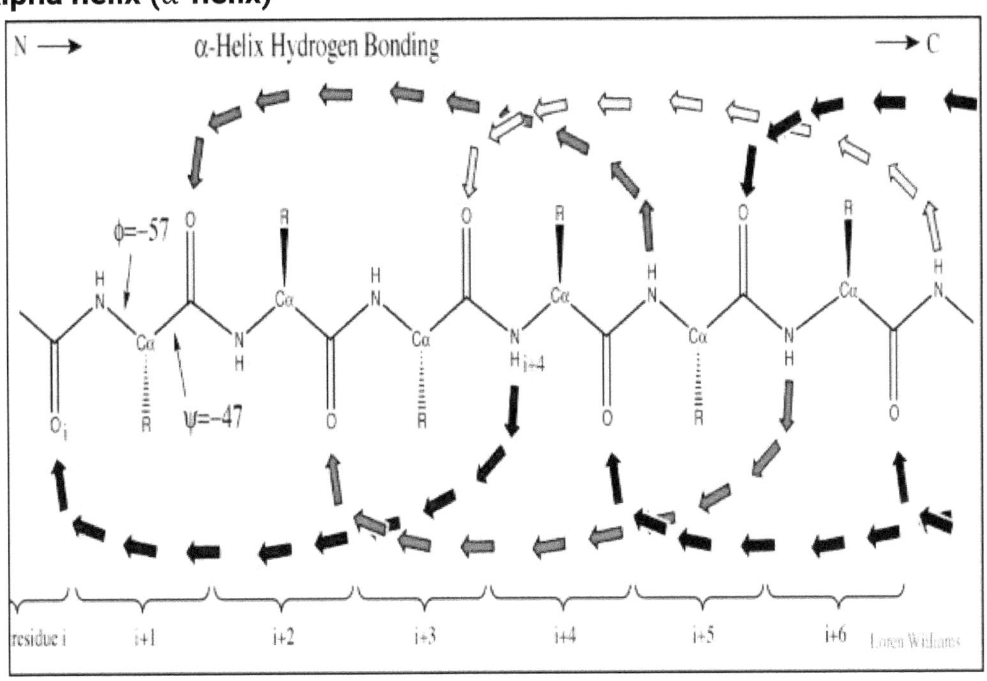

α Helix H-Bonding

The **alpha helix** (α-**helix**) is a common secondary structure of proteins and is a right hand- coiled or spiral conformation (helix) in which every backbone N-H group donates a hydrogen bond to the backbone C=O group of the amino acid four residues earlier ($I + 4 \rightarrow i$ hydrogen bonding). This secondary structure is also sometimes called a classic **Pauling–Corey–Branson alpha helix** (see below). The name **3.613-helix** is also used for this type of helix, denoting the number of residues per helical turn, and 13 atoms being involved in the ring formed by the hydrogen bond. Among types of local structure in proteins, the α- helix is the most regular and the most predictable from sequence, as well as the most prevalent.

PROPERTIES

The amino acids in an α helix are arranged in a right-handed helical structure where each amino acid residue corresponds to a 100° turn in the helix (i.e., the helix has 3.6 residues per turn), and a translation of 1.5 Å (0.15 nm) along the helical axis.

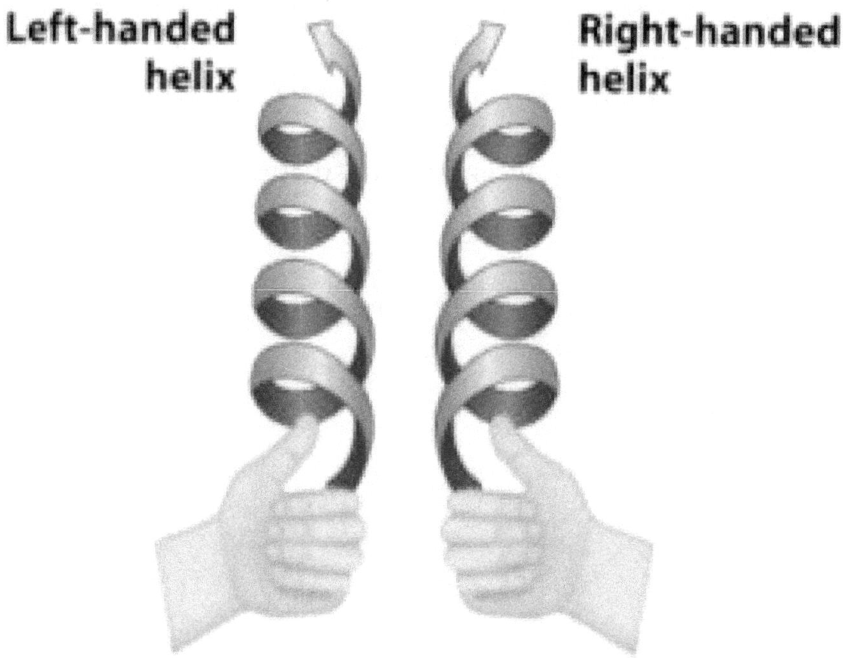

α **Helix – Left handed & Right handed**

Short pieces of left-handed helix sometimes occur with a large content of achiral glycine amino acids, but are unfavorable for the other normal, biological L- amino acids.

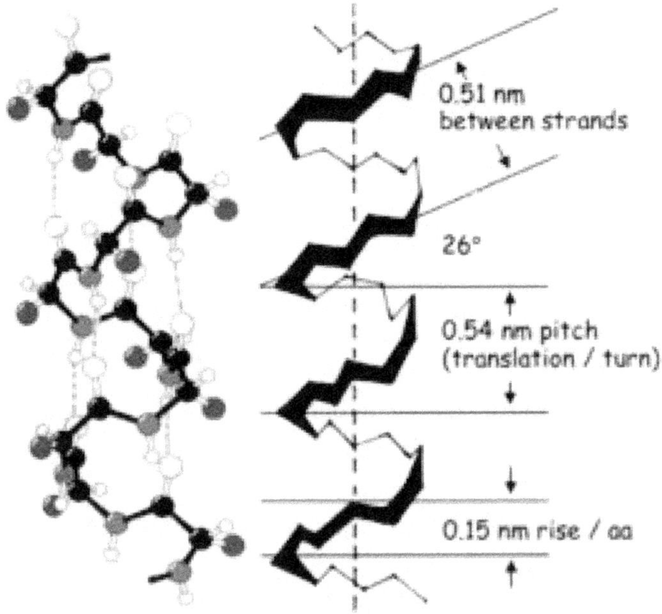

α Helix Pitch

The pitch of the alpha-helix (the vertical distance between consecutive turns of the helix) is

5.4 Å (0.54 nm), which is the product of 1.5 and 3.6. What is most important is that the N- H group of an amino acid forms a hydrogen bond with the C=O group of the amino acid *four* residues earlier; this repeated (*i*+ 4→ *i*) hydrogen bonding is the most prominent characteristic of an α-helix.

Similar structures include the 310 helix (*i*+3→*i* hydrogen bonding) and the π- helix (*i*+5→*i* hydrogen bonding). The α helix can be described as a 3.613 helix, since the i + 4 spacing adds 3 more atoms to the H-bonded loop compared to the tighter 310 helix, and on average, 3.6 amino acids are involved in one ring of α helix. The subscripts refer to the number of atoms (including the hydrogen) in the closed loop formed by the hydrogen bond.

Residues in α-helices typically adopt backbone (φ, ψ) dihedral angles around (-60°, - 45°), as shown in the image at right. In more general terms, they adopt dihedral angles such that the ψ dihedral angle of one residue and the φ dihedral angle of the *next* residue sum to roughly - 105°. As a consequence, α-helical dihedral angles, in general, fall on a diagonal stripe on the Ramachandran diagram (of slope -1), ranging from (-90°, -15°) to (-35°, -70°). For comparison, the sum of the dihedral angles for a 310 helix is roughly -75°, whereas that for the π-helix is roughly -130°.

Table 1- Structural features of the three major forms of protein helices

Geometry attribute	α-helix	310 helix	π-helix
Residues per turn	3.6	3.0	4.4
Translation per residue	1.5 Å (0.15 nm)	2.0 Å (0.20 nm)	1.1 Å (0.11 nm)
Radius of helix	2.3 Å (0.23 nm)	1.9 Å (0.19 nm)	2.8 Å (0.28 nm)
Pitch	5.4 Å (0.54 nm)	6.0 Å (0.60 nm)	4.8 Å (0.48 nm)

Pauling and Corey predicted a second type of repetitive structure, the β conformation -an **extended** state for which angles phi = -135o and psi = +135o; the polypeptide chain **alternates** in direction, resulting in a zig-zag structure for the peptide chain. Note the shaded circle around R; the extended strand arrangement also allows the **maximum space and freedom of movement for a side chain.** The repeat between identically oriented R-groups is 7.0 Å, with 3.5 Å per amino acid, matching the fiber diffraction data for beta-keratins.

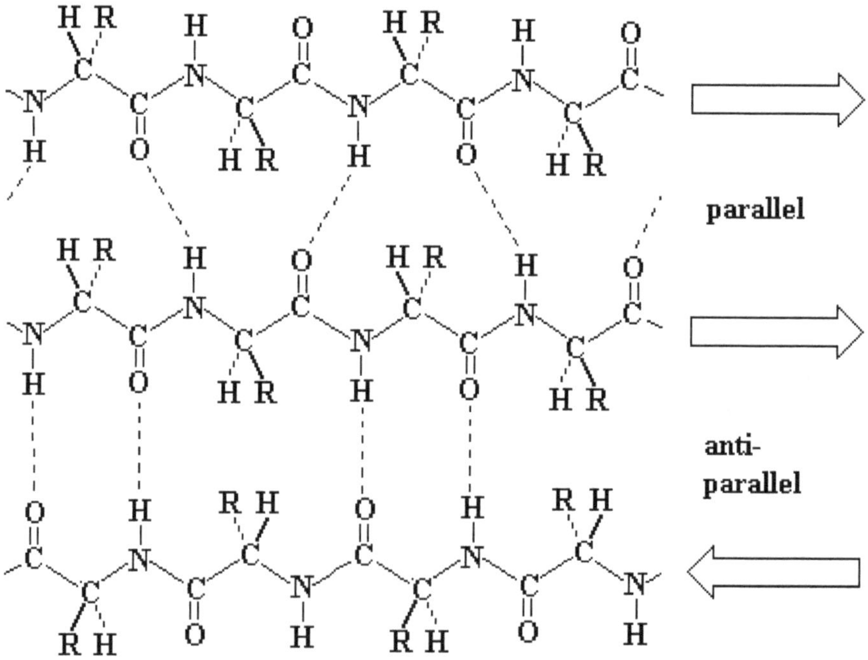

Parallel and Antiparallel beta sheets

Pauling's extended state model matched the spacing of fibroin exactly (3.5 and 7.0 Å). In the extended state, H-bonding NH and CO groups point out at 90o to the strand. If extended strands are lined up side by side, H-bonds bridge from strand to strand. Identical or opposed strand alignments make up parallel or antiparallel beta sheets (named for beta keratin). Antiparallel beta-sheet is significantly more stable due to the well aligned H-bonds.

Table 2 – Dihedral angles in beta sheets

Angle	Antiparallel	Parallel
Φ	-139°	-199°
Ψ	135°	113°

The amino acids have side chains which **disrupt secondary structure**, and are known as **secondary structure breakers:**

- **side chain H is too small** to protect backbone H-bond: **Gly** side chain linked to alpha N, **has no N-H** to H-bond;
- **rigid structure** due to ring restricts to phi = -60o; **Pro**
- **H-bonding side chains** compete directly with backbone H-bonds

Asp, Asn, Ser

Clusters of breakers give rise to regions known as **loops or turns** which mark the boundaries of regular secondary structure, and serve to link up secondary structure segments.

Protein Tertiary Structure

Tertiary structure refers to the three-dimensional arrangement of all atoms in a protein. Tertiary structure is formed by the folding in three dimensions of the secondary structure elements of a protein. While the α helical secondary structure is held together by interactions between the carbonyl and amide groups within the backbone, tertiary structure is held together by interactions between R-groups of residues brought together by folding. Disulfide bonds are also counted under the category of tertiary structure interactions. Proteins that are compact are known as globular proteins.

Examination of protein structures resolved by X-ray diffraction and NMR has revealed a variety of folding patterns common to many different proteins. However, even within these folds, distinct substructures or structural **motifs**, i.e. distinctive arrangements of elements of secondary structure, have been described. The term **supersecondary structure** has been coined to describe this level of organisation, which is intermediate between secondary and tertiary.

Motifs or folds, are particularly stable arrangements of several elements of the secondary structure. Supersecondary structures are usually produced by packing side chains from adjacent secondary structural elements close to each other.

Rules for secondary structure

- Hydrophobic side groups must be buried inside the folds, therefore, layers must be created (β-α-β; α- α).

- α-helix and β-sheet, if occur together, are found in different structural layers.
- Adjacent polypeptide segments are stacked together.
- Connections between secondary structures do not form knots.
- The β-sheet is the most stable.

Motif

- Secondary structure composition, e.g. all α, all β, segregated α+β, mixed α/β
- Motif = small, specific combinations of secondary structure elements, e.g. β-α-β loop

Helix super secondary structures Helix-Turn-Helix Motif

Also called the alpha-alpha type (αα-type). The motif is comprised of two antiparallel helices connected by a turn. The helix-turn-helix is a functional motif and is usually identified in proteins that bind to DNA minor and major grooves, and Calcium-binding proteins.

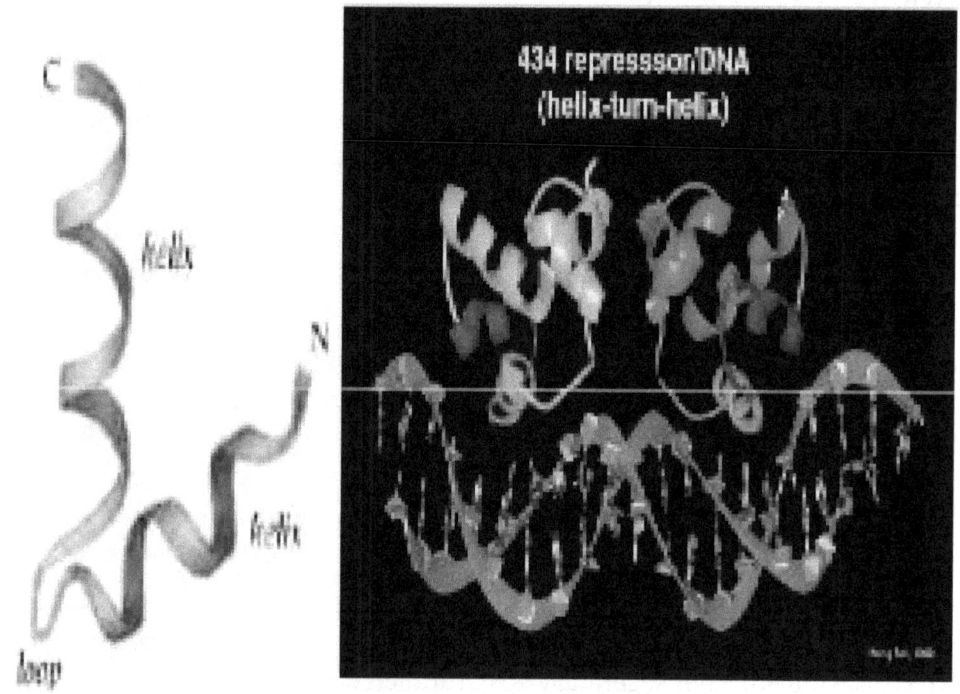

DNA binding Helix-turn-Helix motif

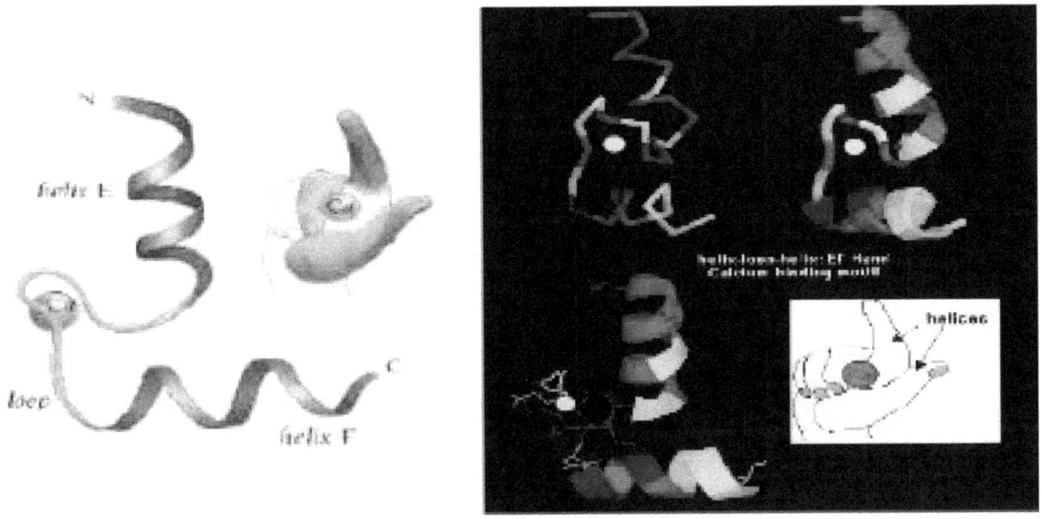

Calcium binding (EF Hand- Calcium binding) motif

Helix-hairpin-helix: Involved in DNA binding

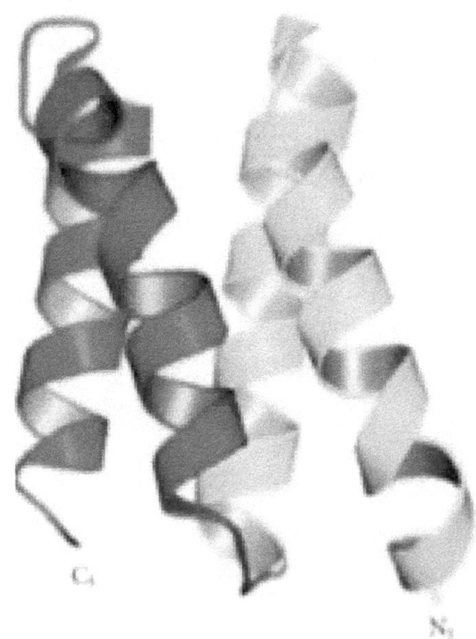

Helix-hairpin-helix

Alpha-alpha corner

Short loop regions connecting helices which are roughly perpendicular to one another

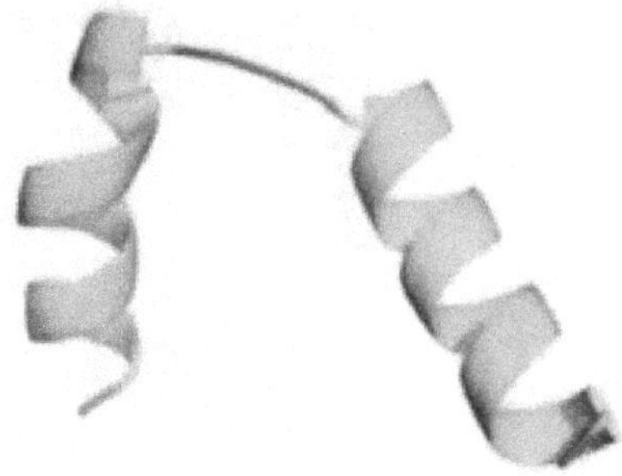

Alpha-alpha corner

Sheet super secondary structures

- All beta tertiary structural domains can occur in proteins with one domain (eg. concanavalin A, superoxide dismutase), and occurs at least once in proteins with two domains (eg. chymotrypsin), or three domains (eg. OmpF).
- The beta strands making up these domains are all essentially antiparallel and form structures to achieve stable packing arrangements within the protein.
- There are presently (as of version 1.39) about 70 subclasses listed in SCOP for this domain, and some examples of these are outlined below.

Beta barrels

This is the most abundant beta-domain structure and as the name suggests the domain forms a 'barrel-like' structure. The beta barrels are not geometrically perfect and can be rather distorted.

There are three main types:

- Up-and-down barrels
- Greek key barrels
- Jelly roll (Swiss roll) barrels

Up-and-down beta-sheets or beta-barrels

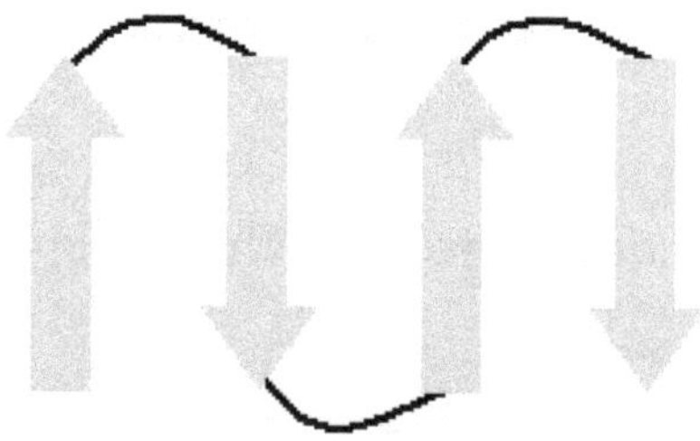

Beta barrels

The simple topology of an up-and-down barrel (named because the beta strands follow each other in sequence in an up-and-down fashion). Usually, the loops joining the beta strands do not crossover the 'ends' of the barrel.

Greek key barrels

These are barrels formed from two, or more, Greek Key motifs. It is a stable structure. The Greek key barrel consists of four anti-parallel Beta strands where one strand changes the topology direction. Hydrogen bonding occurs between strands 1:4, and strands 2:3. Strand 2 then folds over to form the structural motif.

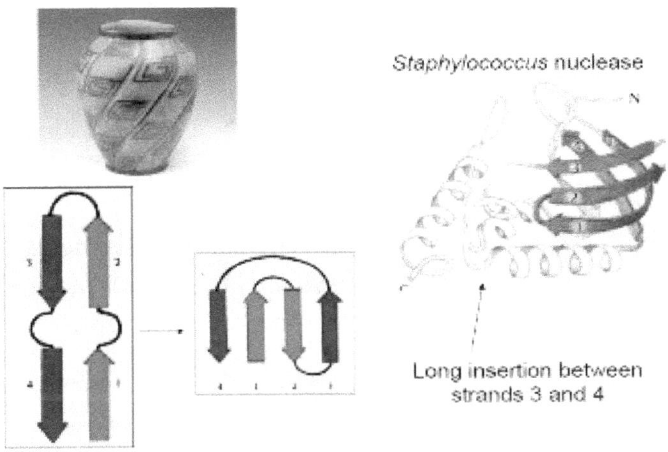

Greek key barrels

Jelly roll barrels

These barrels are formed from a 'Greek Key-like' structure called a jelly roll. Supposedly named because the polypeptide chain is wrapped around a barrel core like a jelly roll (swiss roll). It is a stable structure. This structure is found in coat proteins of spherical viruses, plant lectin concanavalin A, and hemagglutinin protein from influenza virus.

The essential features of a jelly roll barrel are that:

- it is like an inverted 'U' (which is often seen twisted and distorted in proteins) it is usually divided into two beta sheets which are packed against each other most jelly roll barrels have eight strands although any even number greater than 8 can form a jelly roll barrel it folds such that hydrogen bonds exist between strands 1 and 8; 2 and 7; 3 and 6; and 4 and 5

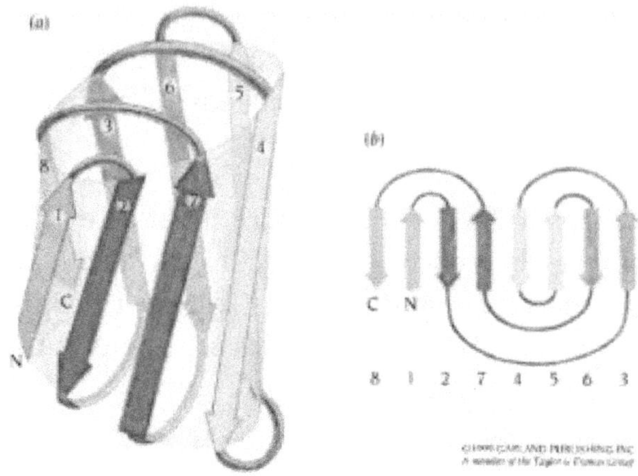

Jelly roll barrels

Beta sandwich

A beta sandwich is essentially a 'flattened' beta barrel with the two sheets packing closely together (like a sandwich!). The first and last strands of the sandwich do not hydrogen bond to each other to complete a 'barrel' structure.

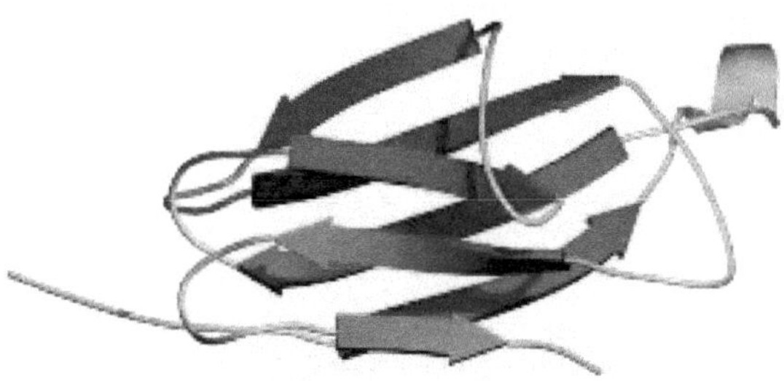

The structure of human beta-2-microglobulin

Beta sandwich in beta 2 microglobulin

Aligned or Orthogonal beta strands

Beta strands in barrels or sandwich structures can be orientated in two general ways: where the strands in two sheets are almost aligned, and in the same orientation, to each other and form an 'aligned beta' structure (eg. gamma crystallin) where the strands, in at least two sheets, are roughly perpendicular to each other and form an 'orthogonal beta' structure.

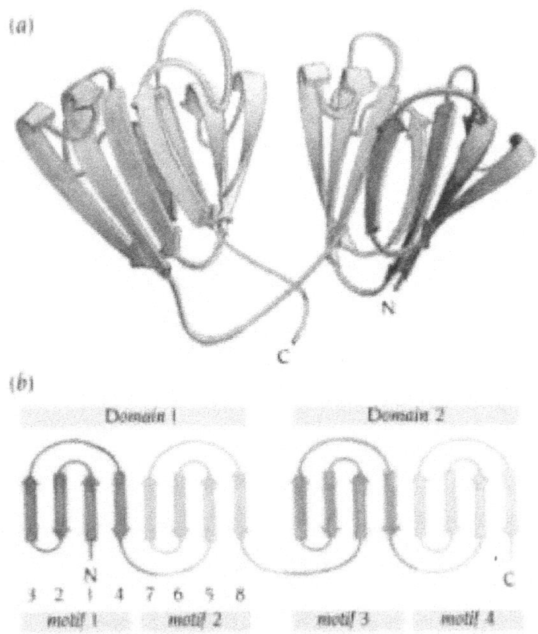

Aligned beta strands

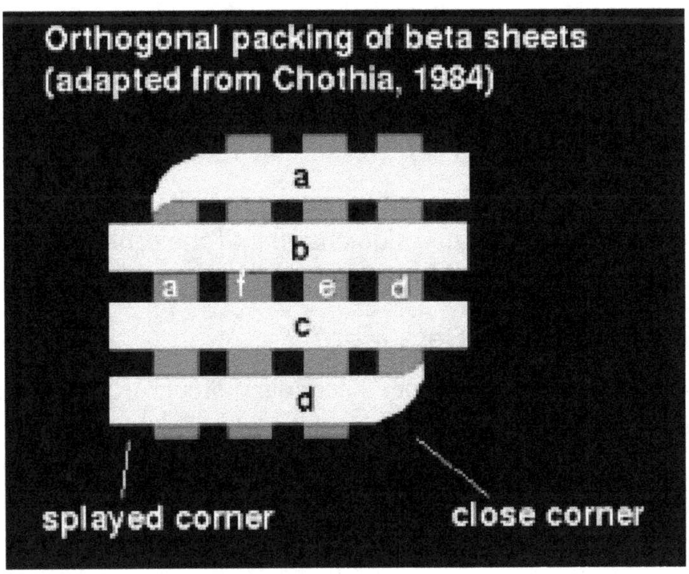

Orthogonal beta sheets

Beta-hairpin: two antiparallel beta strands connected by a "hairpin" bend, i.e. beta-turn 2 x antiparallel beta-strands + beta-turn = beta hairpin

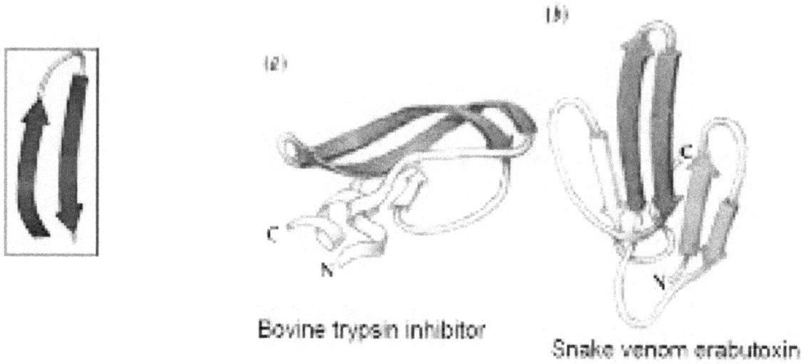

Beta hairpin

Beta-beta corner

Beta-beta corner

Two antiparallel beta strands which form a beta hairpin can change direction abruptly. The angle of the change of direction is about 90 degrees and so the structure is known as a 'beta corner'

The abrupt angle change is achieved by one strand having a glycine residue (so there is no steric hindrance from a side chain) and the other strand having a beta bulge (where the hydrogen bond is broken).

α/β Topologies Beta-Helix-Beta Motif

An important and widespread super secondary structural motif in proteins is known as the β- α-β motif (Beta-Alpha-Beta motif). The motif consists of two parallel Beta strands that is connected via an alpha helix (with two turns). The motif is found in most proteins that contain parallel beta strands, and the axis of the Helix and the Strands are roughly parallel to each other with all three elements forming a hydrophobic core due to shielding. The β-α-β motif may be structurally or functionally involved. The Loop that connects the C-terminal of first Beta strand and N-terminal of Helix is frequently involved in ligand binding functions, and the motif itself is frequently found in ion channels.

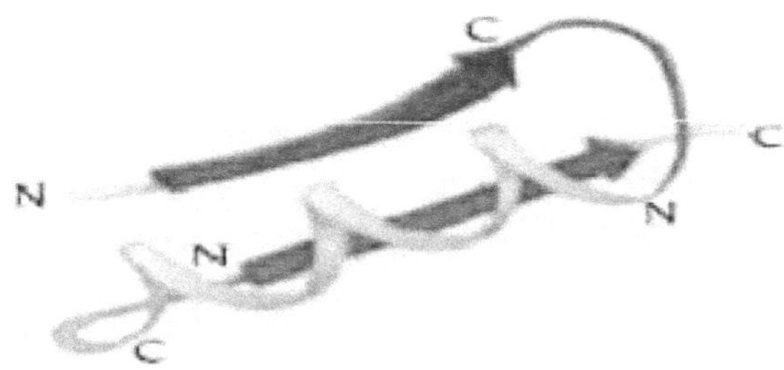

Beta-Helix-Beta Motif

The β - α - β - α - β subunit, often present in nucleotide-binding proteins, is named the **Rossman Fold**, after Michael Rossman

The Rossman fold

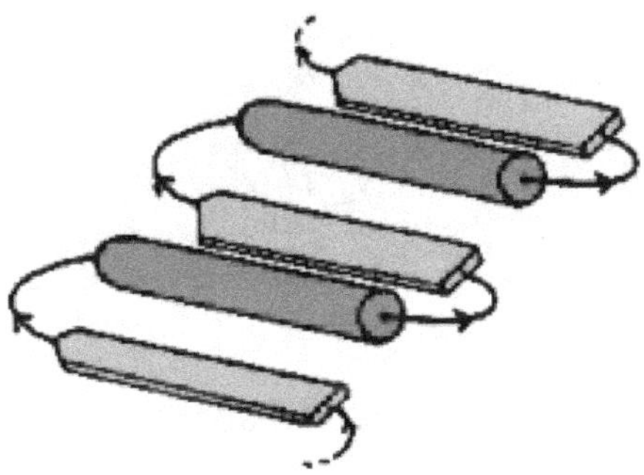

Rossman fold

α/β horse shoe

- stranded parallel b sheet curved into an open horseshoe shape, with 16 a-helices packed against the outer surface. It doesn't form a barrel although it looks as though it should. The strands are only very slightly slanted, being nearly parallel to the central `axis'.

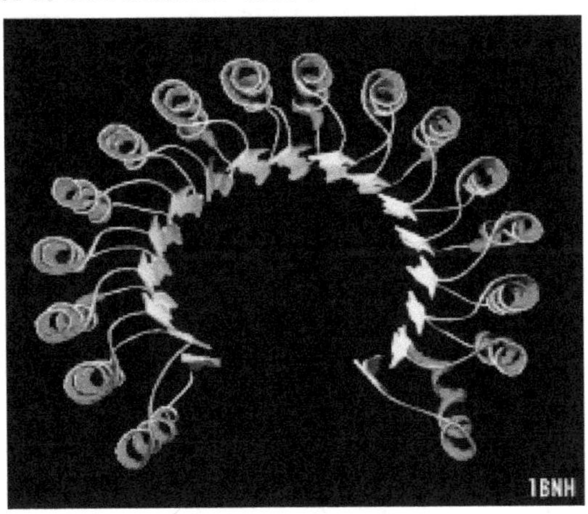

α/β **horse shoe**

α/β barrels

Consider a sequence of eight α/β motifs:

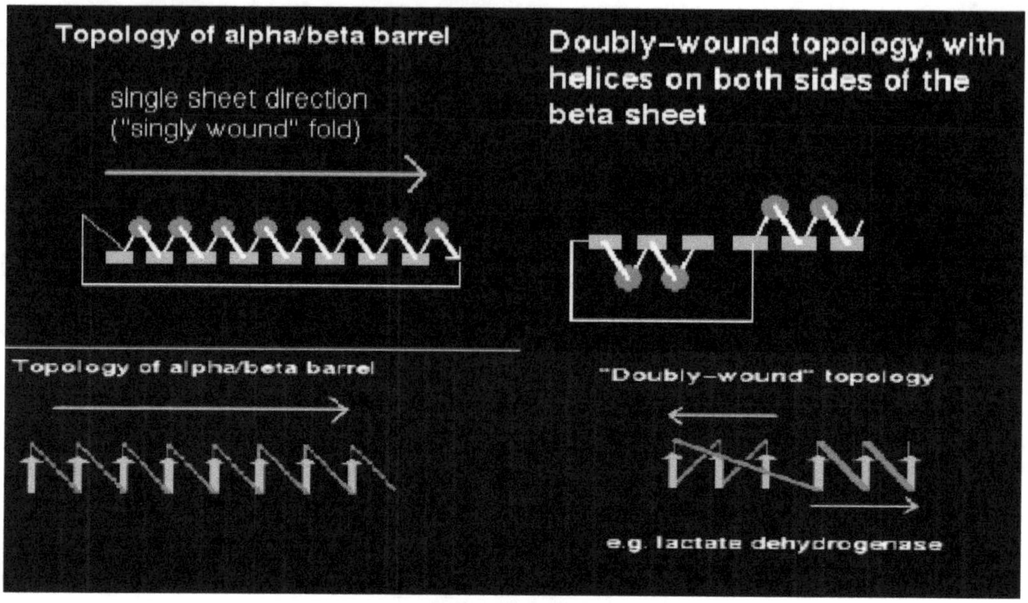

Topology of α/β barrel

If the first strand hydrogen bonds to the last, then the structure closes on itself forming a barrel-like structure. This is shown in the picture of triose phosphate isomerase.

Note that the "staves" of the barrel are slanted, due to the twist of the b sheet. Also notice that there are effectively four layers to this structure. The direction of the sheet does not change (it is anticlockwise in the diagram). Such a structure may therefore be described as **singly wound**.

In a structure which is open rather than closed like the barrel, helices would be situated on only one side of the b sheet if the sheet direction did not reverse. Therefore open a/b structures must be **doubly wound** to cover both sides of the sheet.

The chain starts in the middle of the sheet and travels outwards, then returns to the centre via a loop and travels outwards to the opposite edge:

Doubly-wound topologies where the sheet begins at the edge and works inwards are rarely observed.

Alpha+Beta Topologies

This is where we collect together all those folds which include significant alpha and beta secondary structural elements, but for which those elements are **`mixed'**, in the sense that they do NOT exhibit the wound alpha-beta topology. This class of folds is therefore referred to as α+ β

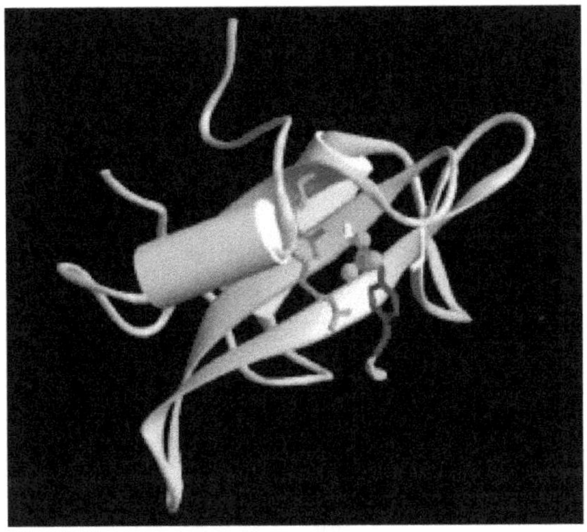

Alpha+Beta Topology

Domains

Domains are stable, independently folded, globular units, often consisting of combinations of motifs vary from 25 to 300 amino acids, average length – 100. large globular proteins may consist of several domains linked by stretches of polypeptide. Separate domain may have distinct functions (eg G3P dehydrogenase). In many cases binding site formed by cleft between 2 domains frequently correspond to exon in gene. Some examples of domains:

- involving α-helix 4-helix bundle globin fold

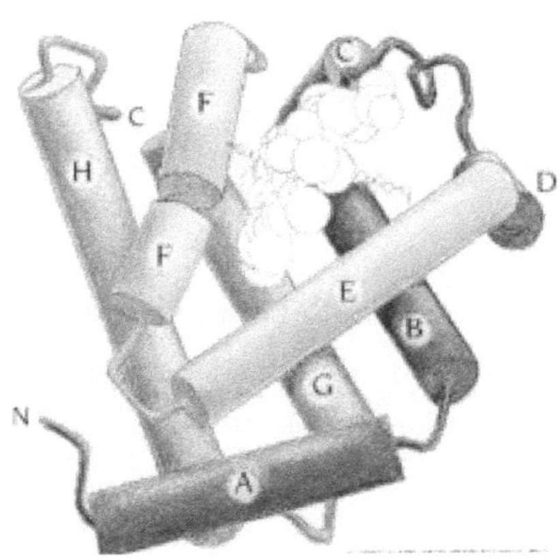

α-**helix 4-helix bundle globin fold**

The globin fold is found in its namesake globin protein families: hemoglobins and myoglobins, as well as in phycocyanins. Because myoglobin was the first protein whose structure was solved, the globin fold was thus the first protein fold discovered.

parallel β-sheets

hydrophobic residues on both sides, therefore must be buried.

barrel: 8 β strands each flanked by an antiparallel α-helix eg triose phosphate isomerase.)

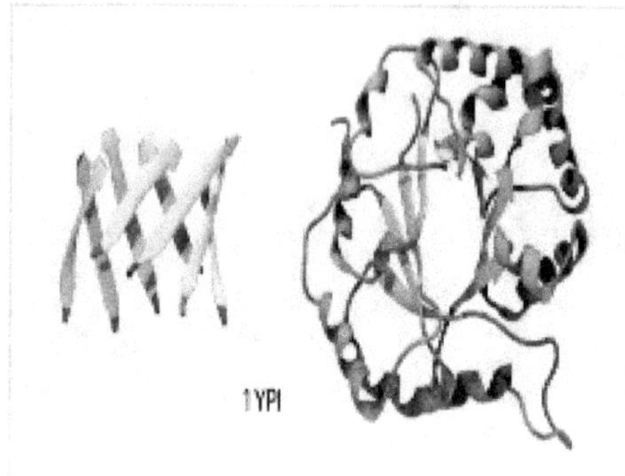

Parallel beta sheets

antiparallel β -sheet

Hydrophobic residues on one side, one side can be exposed to environment, minimum structure 2 layers Sheets arranged in a barrel shape. More common than parallel β - barrels eg. immunoglobulin

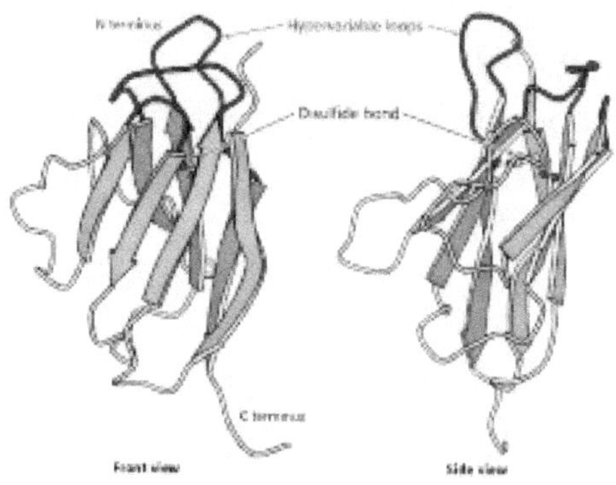

Antiparallel beta sheets

The **immunoglobulin domain** is a type of protein domain that consists of a 2-layer sandwich of 7-9 antiparallel β-strands arranged in two β-sheets with a Greek key topology, consisting of about 80 amino acids.

The backbone switches repeatedly between the two β-sheets. Typically, the pattern is (N- terminal β-hairpin in sheet 1)-(β-hairpin in sheet 2)-(β-strand in sheet 1)-(C-terminal β- hairpin in sheet 2). The cross-overs between sheets form an «X», so that the N- and C- terminal hairpins are facing each other.

Members of the immunoglobulin superfamily are found in hundreds of proteins of different functions. Examples include antibodies, the giant muscle kinase titin, andreceptor tyrosine kinases. Immunoglobulin-like domains may be involved in protein– protein and protein– ligand interactions.

Example of Tertiary Structure: Myoglobin and Hemoglobin

Myoglobin and hemoglobin are hemeproteins whose physiological importance is principally related to their ability to bind molecular oxygen.

Myoglobin

Single polypeptide chain (153 amino acids).

No disulfide bonds 8 right handed alpha helices form a hydrophobic pocket which contains heme molecule protective sheath for a heme group

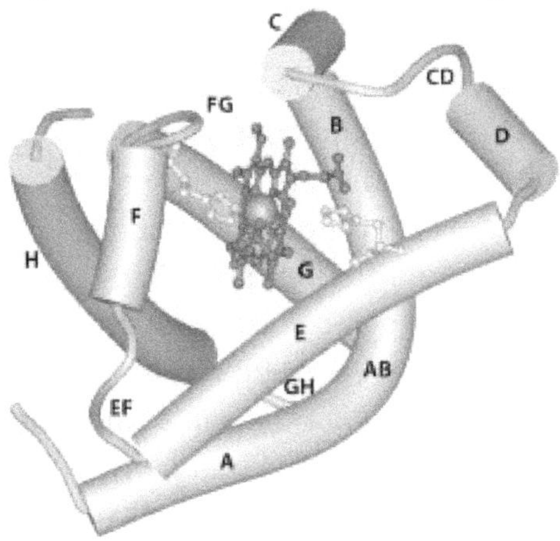

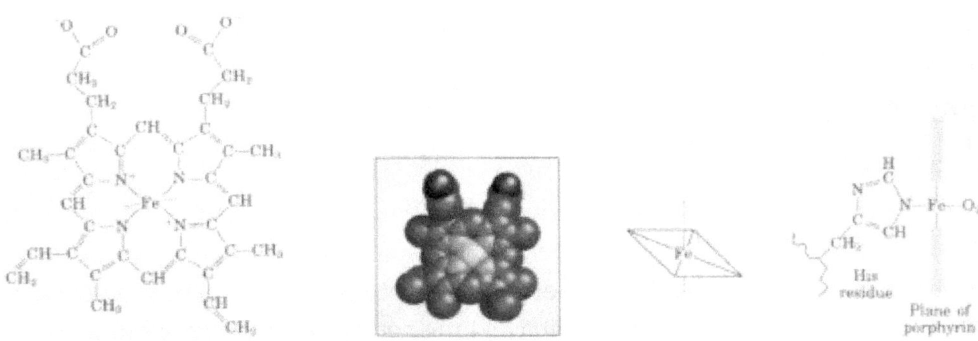

Myoglobin structure

Myoglobin is a monomeric heme protein found mainly in muscle tissue where it serves as an intracellular storage site for oxygen During periods of oxygen deprivation oxymyoglobin releases its bound oxygen which is then used for metabolic purposes The tertiary structure of myoglobin is that of a typical water soluble globular protein Its secondary structure is unusual in that it contains a very high proportion (75%) of α-helical secondary structure A myoglobin polypeptide is comprised of 8 separate right handed a-helices, designated A through H, that are connected by short non helical regions Amino acid R-groups packed into the interior of the molecule are predominantly hydrophobic in character while those exposed on the surface of the molecule are generally hydrophilic, thus making the molecule relatively water soluble.

Each myoglobin molecule contains one heme prosthetic group inserted into a hydrophobic cleft in the protein Each heme residue contains one central coordinately bound iron atom that is normally in the Fe 2+ , or ferrous, oxidation state The oxygen carried by hemeproteins is bound directly to the ferrous iron atom of the heme prosthetic group. The heme group is located in a crevice Except for one edge, non polar side chains surround the heme Fe 2+ is octahedrally coordinated Fe 2+ covalently bonded to the imidazole group of histidine 93 (F8) O2 held on the other side by histidine 64 (E7). Hydrophobic interactions between the tetrapyrrole ring and hydrophobic amino acid R groups on the interior of the cleft in the protein strongly stabilize the heme protein conjugate. In addition a nitrogen atom from a histidine R group located above the plane of the heme ring is coordinated with the iron atom further stabilizing the interaction between the heme and the protein. In oxymyoglobin the remaining bonding site on the iron atom (the 6th coordinate position) is occupied by the oxygen, whose binding is stabilized by a second histidine residue Carbon monoxide also binds coordinately to heme iron atoms in a manner similar to that of oxygen, but the binding of carbon monoxide to heme is much stronger than that of oxygen. The preferential binding of carbon monoxide to heme iron is largely responsible for the asphyxiation that results from carbon monoxide poisoning.

Hemoglobin

Oxygen transporter Four polypeptide chains Tetramer Each chain has a heme group Hence four O 2 can bind to each Hb Two alpha (141 amino acids) and two beta (146 amino acids) chains

Hemoglobin is an [α(2):β(2)] tetrameric hemeprotein found in erythrocytes where it is responsible for binding oxygen in the lung and transporting the bound oxygen throughout the body where it is used in aerobic metabolic pathways Each subunit of a hemoglobin tetramer has a heme prosthetic group identical to that described for myoglobin. Although the secondary and tertiary structure of various hemoglobin subunits are similar, reflecting extensive homology in amino acid composition, the variations in amino acid composition that do exist impart marked differences in hemoglobin's oxygen carrying properties In addition, the quaternary structure of hemoglobin leads to physiologically important allosteric interactions between the subunits, a property lacking in monomeric myoglobin which is otherwise very similar to the α-subunit of hemoglobin

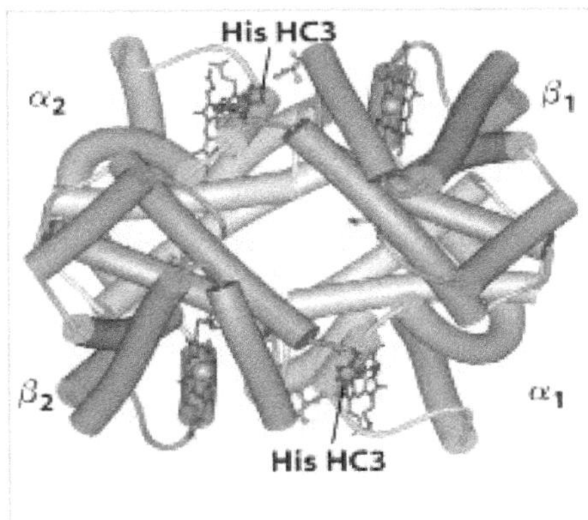

Hemoglobin structure

Quaternary structure

3-dimensional relationship of the different polypeptide chains (subunits) in a multimeric protein, the way the subunits fit together and their symmetry relationships.

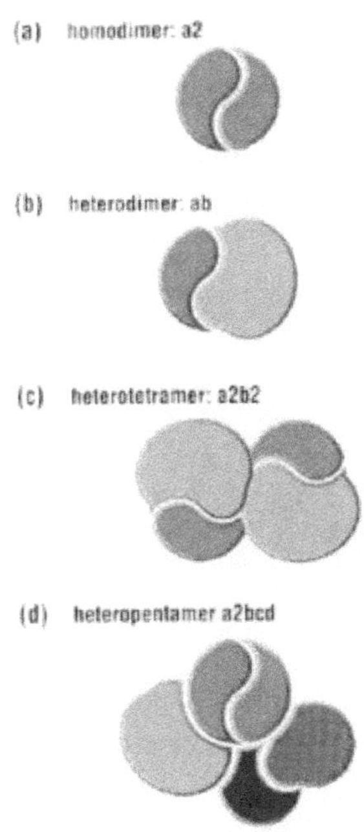

Protein subunits

- NOTE: This use of the Greek letters to differentiate different polypeptide chains in a multimeric protein has nothing to do with the names for the secondary structures α helix and β conformation.

- Only in proteins with more than one polypeptide chain; proteins with only one chain have no quaternary structure.
- Each polypeptide chain in a multichain protein = a subunit
- 2-subunit protein = a dimer, 3 subunits = trimeric protein, 4 = tetrameric
- homo(dimer or trimer etc.): identical subunits
- hetero(dimer or trimer etc.): more than one kind of subunit (chains with different amino acid sequences)
- different subunits designated with Greek letters – e.g., subunits of a heterodimeric protein = the "α subunit" and the "β subunit".
- Some protein structures have very complex quaternary arrangements; e.g., mitochondrial ATP synthase, viral capsids

Symmetry in quaternary structures

- Simplest kind of symmetry = rotational symmetry
- Individual subunits can be superimposed on other identical subunits (brought into coincidence) by rotation about one or more rotational axes.
- If the required rotation = 180° (360°/2), protein has a 2-fold axis of symmetry (e.g., Cro repressor protein above).
- If the rotation = 120° (360°/3), e.g., for a homotrimer, the protein has a 3-fold symmetry axis. Rotational symmetry in proteins: Cyclic symmetry: all subunits are related by rotation about a single n-fold rotation axis (C2 symmetry has a 2-fold axis, 2 identical subunits; C3 symmetry has a 3-fold axis, 3 identical subunits, etc.)

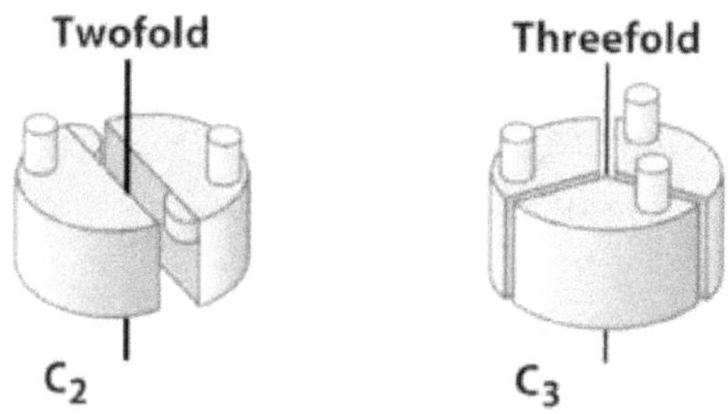

Two common folds of symmetry

Forces that stabilize Protein Structure

Proteins are formed of amino acids linked together by the following types of bonds

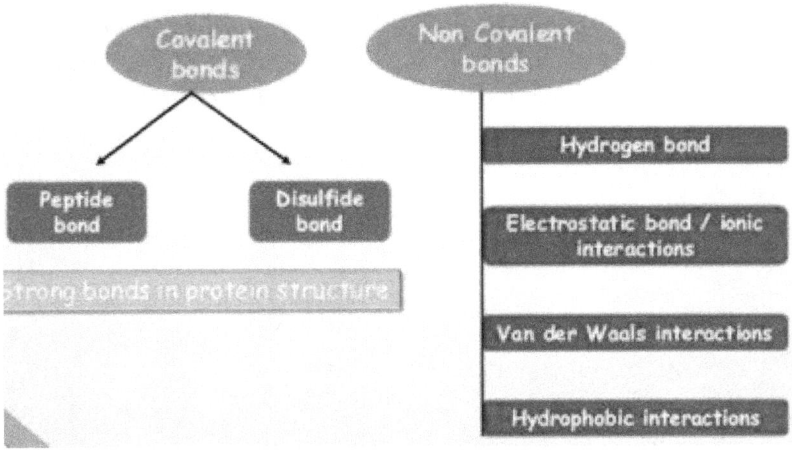

Forces stabilizing protein structure

Covalent Bonds - Disulfide Bridges

Covalent bonds are the strongest chemical bonds contributing to protein structure. Covalent bonds arise when two atoms share electrons.

In addition to the covalent bonds that connect the atoms of a single amino acid and the covalent peptide bond that links amino acids in a protein chain, covalent bonds between

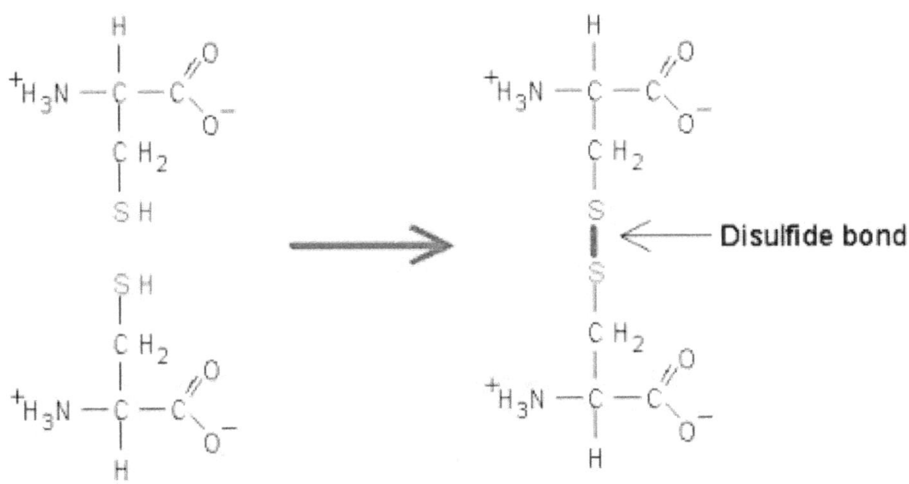

Disulphide bond

Non-covalent bonds Electrostatic Interactions Ionic Bonds - Salt Bridges

Ionic bonds are formed as amino acids bearing opposite electrical charges are juxtaposed in the hydrophobic core of proteins. Ionic bonding in the interior is rare because most charged amino acids lie on the protein surface. Although rare, ionic bonds can be important to protein structure because they are potent electrostatic attractions that can approach the strength of covalent bonds. An ionic bond-salt bridge between a negatively charged O on the sidechain of glutamic acid lies 2.8 Å from the positively charged N on the amino terminus (lysine) is shown here.

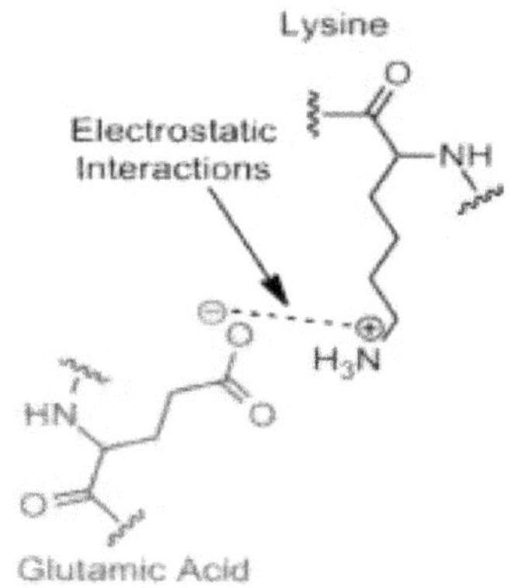

Electrostatic bond

Hydrogen Bonds

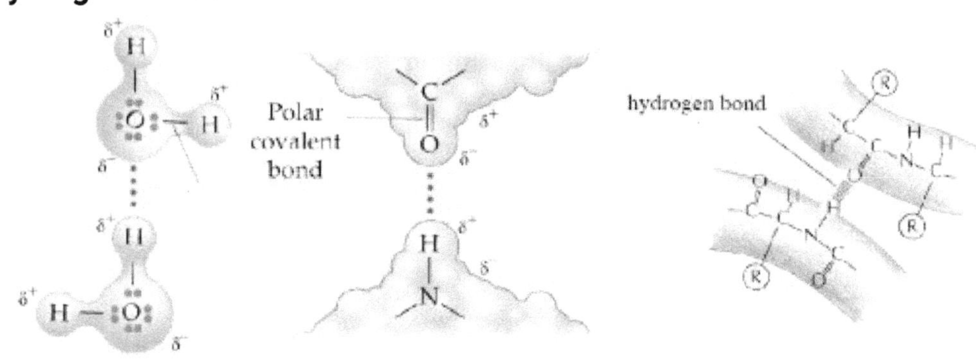

Two water molecules Two parts of one large molecule

Hydrogen bond

Hydrogen bonds are a particularly strong form of dipole-dipole interaction. Because atoms of different elements differ in their tendencies to hold onto electrons -- that is, because they have different electronegativities -- all bonds between unlike atoms are polarized, with more electron density residing on the more electronegative atom of the bonded pair. Separation of partial charges creates a dipole, which you can think of as a mini-magnet with a positive and a negative end. In any system, dipoles will tend to align so that the positive end of one dipole and the negative end of another dipole are in close proximity. This alignment is favorable.

Hydrogen bonds are dipole-dipole interactions that form between heteroatoms in which one heteroatom (e.g. nitrogen) contains a bond to hydrogen and the other(e.g.

oxygen) contains an available lone pair of electrons. You can think of the hydrogen in a hydrogen bond as being shared between the two heteroatoms, which is highly favorable. Hydrogen bonds have an ideal X-H-X angle of 180°, and the shorter they are, the stronger they are. Hydrogen bonds play an important role in the formation of secondary structure. Alpha helices are hydrogen bonded internally along the backbone whereas beta strands are hydrogen bonded to other beta strands. Side chains can also participate in hydrogen bonding interactions. You should be able to list the side chains that can participate in hydrogen bonds now that you know the structures of the side chains. Because hydrogen bonds are directional, meaning the participating dipoles must be aligned properly for a hydrogen bond to form (another w ay of saying it is that the hydrogen bonding angle must be larger than about 135°, with an optimum of 180°), and because unfavorable alignment of participating dipoles is repulsive, hydrogen bonds between side chains play key roles in determining the unique structures that different proteins form.

Hydrophobic Bonds

Hydrophobic bonds are a major force driving proper protein folding. Burying the nonpolar surfaces in the interior of a protein creates a situation where the water molecules can hydrogen bond with each other without becoming excessively ordered. Thus, the energy of the system goes down.

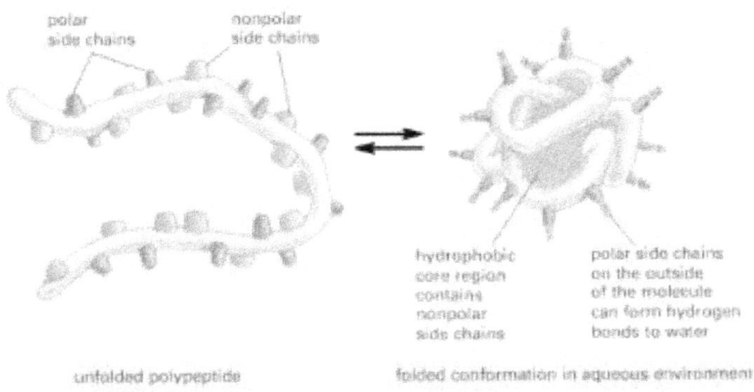

Hydrophobic bonds

Therefore, an important factor governing the folding of any protein is the distribution of its polar and nonpolar amino acids. The nonpolar (hydrophobic) side chains in a protein such as those belonging to phenylalanine, leucine, isoleucine, valine, methionine and tryptophan tend to cluster in the interior of the molecule (just as hydrophobic oil droplets coalesce in water to form one large droplet). In contrast, polar side chains such as those belonging to arginine, glutamine, glutamate, lysine, etc. tend to arrange themselves near the outside of the molecule, where they can form hydrogen bonds with water and with other polar molecules. There are some polar amino acids in protein interiors, however, and these are very important in defining the precise shape adopted by the protein because the pairing of opposite poles is even more significant than it is in water.

Van der Waals Forces

The Van der Waals force is a transient, weak electrical attraction of one atom for another. Van der Waals attractions exist because every atom has an electron cloud that can fluctuate, yielding a temporary electric dipole. The transient dipole in one atom can induce a complementary dipole in another atom, provided the two atoms are quite close. These short- lived, complementary dipoles provide a weak electrostatic attraction, the Van der Waals force. Of course, if the two electron clouds of adjacent atoms are too close, repulsive forces come into play because of the negatively-charged electrons. The appropriate distance required for Van der Waals attractions differs from atom to atom, based on the size of each electron cloud, and is referred to as the Van der Waals radius. The dots around atoms in this and other displays represent Van der Waals radii.

Van der Waals attractions, although transient and weak, can provide an important component of protein structure because of their sheer number. Most atoms of a protein are packed sufficiently close to others to be involved in transient Van der Waals attractions.

Van der Waals forces can play important roles in protein-protein recognition when complementary shapes are involved. This is the case in antibody-antigen recognition, where a "lock and key" fit of the two molecules yields extensive Van der Waals attractions.

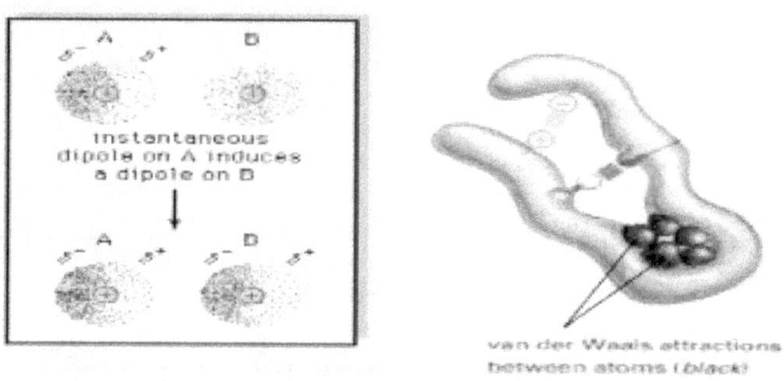

Van der Waals forces

PROTEIN STRUCTURE PREDICTION

Proteins are one of the major biological macromolecules performing a variety functions such as enzymatic catalysis, transport, regulation of metabolism, nerve conduction, immune response etc. The three-dimensional structure of a protein is responsible for its function.

Sequence-Structure Gap and the Need for Structure Prediction

With the advent of recombinant DNA technology it has become possible to determine the amino acid sequences of proteins quite rapidly. However, determining the three dimensional structure of proteins is a time consuming task and hence there exists a vast gap between the number of proteins of known amino acid sequence and that of known structures. This is called as the sequence-structure gap. As the knowledge of the 3-D structure of a protein is very essential to understand its function, it is imperative to develop techniques to predict the structure of a protein from its amino acid sequence.

Basis for Structure Prediction:

The classic experiments carried out by C.B. Afinson in the 60's on the enzyme ribonuclease led to the conclusion that the information to specify the 3-D structure of a protein resides in its amino acid sequence. Within the cell a newly synthesized protein chain spontaneously folds into the compact globular structure to perform its function. Thus nature has an algorithm to fold proteins to their native structures. Efforts have been directed for the past four decades to discover nature's algorithm and computational methods have been developed to predict the structure of proteins from their sequences.

Approaches to Structure Prediction

Prediction of protein structures can be classified into two major categories viz.

1. Prediction of secondary structure and
2. Prediction of tertiary (3-D) structure.

Prediction of secondary structure of proteins attempts to locate segments of the polypeptide

chain adopting the α-helical or β-strand structure. Regions that are devoid of these regular secondary structural elements are considered to adopt coil conformation.

In tertiary structure prediction, one attempts to predict the three-dimensional structure of a protein or the native structure. While so far this has remained an elusive goal, different methods have been developed to press forward to the attainment of this goal.

Secondary structure prediction What?

- Given a protein sequence (primary structure)

GHWIATRGQLIREAYEDYRHFSSECPFIP

(C=Coils H=Alpha Helix E=Beta Strands)

CEEEEECHHHHHHHHHHCCCHHCCCCCC

- 1 st step in prediction of protein structure.
- Technique concerned with determination of secondary structure of given polypeptide by locating the Coils Alpha Helix Beta Strands in polypeptide

Why?

- secondary structure —tertiary structure prediction
- Protein function prediction
- Protein classification
- Predicting structural change
- detection and alignment of remote homology between proteins
- on detecting transmembrane regions, solvent-accessible residues, and other important features of molecules
- Detection of hydrophobic region and hydrophilic region

Prediction methods Chou-Fasman method

- Based on the propensities of different amino acids to adopt different secondary structures
- Predictions are made using a rules-based approach to identify groups of amino acids with shared secondary structure propensities
- Garnier, Osguthorpe, Robson (GOR) method

Statistical method of secondary structure prediction based on information theory & Bayesian probability

Multiple Sequence Alignment (MSA) methods

Performs secondary structure prediction on a multiple sequence alignment as opposed to a single protein sequence

Neural network-based methods

Example: Profile network from **Heid**elberg (PHD)

Chou-Fasman method:

Alpha Helix Prediction:

- Nucleate a helix by scanning for groups of 6 residues with at least 4 helix formers (Hα and hα) and no more than 1 helix breaker (Bα and bα).
- Two Iα residues count as one helix former for nucleating a helix
  - Propagate predicted helix in both directions until reach a four residue window with average propensity (Pα) < 1.0
  - The average propensity (Pα) for a predicted helix must be Pα > 1.03 and Pα > Pβ

Beta Strand Prediction:

- Nucleate a β-strand by scanning for groups of 5 residues with at least 3 strand formers (Hβ and hβ) and no more than 1 strand breaker (B$ and b$).
- Propagate predicted β-strand in both directions until reach a four residue window with average propensity (Pβ) < 1.0
- The average propensity (Pβ) for a predicted β-strand must be Pβ > 1.05 and Pβ > Pα

Resolving conflicting predictions:

(regions with both α-helix and β-strand assignment)

- If average Pα > average Pβ, then the region is alpha helix
- If average Pβ > average Pα, then the region is beta strand

Chou-Farman algorithm:

- Later versions of the algorithm included predictions for turns
- The original algorithm contained additional rules about the location of certain residues (e.g., proline) in α-helices and β-strands
- More recent versions of the algorithm have used sequential tetrapeptide average propensities to predict secondary structure
- The propensity values have also been variously recalculated with larger protein data sets

§ Example of Chou-Fasman method:

Sequence: **MLNPKSYENAIQLGRCFTTHYA**

alpha helix nucleation

M	L	N	P	K	S	Y	E	N	A	I	Q	L	G	R	C	F	T	T	H	Y	A
h	h	b	b	h	i	b	h	b	h	h	h	h	b	i	i	h	i	i	I	b	h

- Has at least 4 helix formers
- Has no more than 1 helix breaker

- Note: Counts as 0.5 helix former

propagating alpha helix

Propagate helix in both directions until reach a four residue window with average propensity (P_{α}) < 1.0

GOR (Garnier, Osguthorpe, Robson) Method

Key difference: Chou-Fasman uses individual amino acid propensities, while GOR incorporates information about neighboring amino acids to make prediction

A 20 x 17 matrix of directional information values for each secondary structure class was calculated from a database of known structures

These matrices are used to predict the secondary structure of the central (9th) residue in a 17 residue window:

M L N P K S Y E N A I Q L G R C F T T H Y A

2' structure prediction for residue I is based on 17 residue window

The secondary structure class with highest information score over 17 residue window is selected as the prediction for the central residue of the window (e.g., I is predicted to be α- helix)

Multiple sequence alignment method

A multiple sequence alignment arranges protein sequences into a rectangular array with the goal that residues in a given column are homologous (derived from a common ancestor), superposable (in a 3D structural alignment - α helix / β sheet) or play a common functional role (catalytic sites, nuclear localisation signal, protein-protein interaction sites,...). Uses BLAST to identify homologous protein sequence fragments in a protein structure database (PDB).

```
VTISCTGSSSNIGAG-NHVKWYQQLPG
VTISCTGTS-NIGS-ITVNWYQQLPG-
LRLS-CSVSGFIFSS-YAMYWVRQAPG
-LS-LTCTVSGTSFDDYYSTWVRQPPG
PEVTCVVVDVSHEDPQVKFN-WYVDG-
A--TLVCTISDFYPGAVTVA-WKADS-
AALGCTVKDYFPEPVTVSWN--SG---
VSLTCTVKGFYPSD--IAVEWESNG--
```

Goal: try to have a maximum of identical/similar residues in a given column of the alignment

```
VTISCTGSSSNIGAG-NHVKWYQQLPG
VTISCTGTSSNIGS--ITVNWYQQLPG
LRLSCSVSGFIFSS--YAMYWVRQAPG
LSLTCTVSGTSFDD--YYSTWVRQPPG
PEVTCVVVDVSHEDPQVKFNWYVDG--
ATLVCTISDFYPGA--VTVAWKADS--
AALGCTVKDYFPEP--VTVSWNSG---
VSLTCTVKGFYPSD--IAVEWESNG--
```

Table 1

Protein Tertiary Structures: Prediction from Amino Acid Sequences

The biological function of a protein is often intimately dependent upon its tertiary structure. X-ray crystallography and nuclear magnetic resonance are the two most mature experimental methods used to provide detailed information about protein structures. However, to date the majority of the proteins still do not have experimentally determined structures available. As at December 2000, there were about 14 000 structures available in the protein data bank, and there are about 10 106 000 sequence records sequences in GenBank (http://www.ncbi.nlm.nih.gov/Genbank). Thus theoretical methods are very important tools to help biologists obtain protein structure information. The goal of theoretical research is not only to predict the structures of proteins but also to understand how protein molecules fold into the native structures. The current methods for protein structure prediction can be roughly divided into three major categories: comparative modelling; threading; and ab initio prediction. For a given target protein with unknown structure, the general procedure for predicting its structure is described below:

Comparative modelling

It is based on two major observations:

- The structure of a protein is uniquely determined by its amino acid sequence. Knowing the sequence should, at least in theory, suffice to obtain the structure.

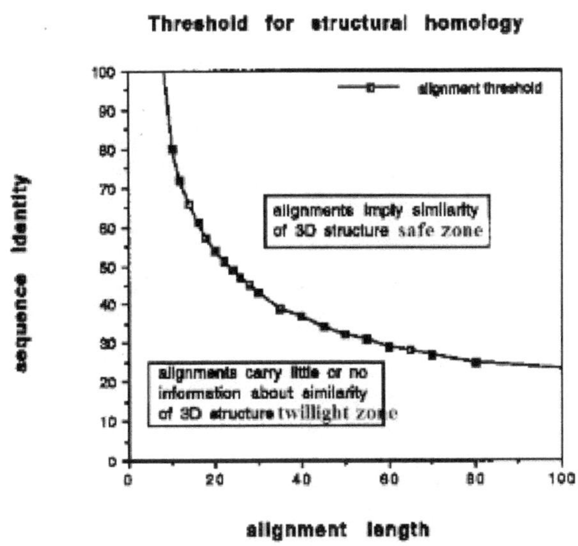

- During evolution, the structure is more stable and changes much slower than the associated sequence, so that similar sequences adopt practically identical structures, and distantly related sequences still fold into similar structures. This relationship was first identified by Chothia and Lesk (1986) and later quantified by Sander and Schneider (1991).

- Thanks to the exponential growth of the Protein Data Bank (PDB), Rost (1999) could recently derive a precise limit for this rule, shown in Figure below. As long as the length of two sequences and the percentage of identical residues fall in the region marked as "safe," the two sequences are practically guaranteed to adopt a similar structure

For a sequence of 100 residues, for example, a sequence identity of 40% is sufficient for structure prediction. When the sequence identity falls in the safe homology modeling zone, we can assume that the 3D-structure of both sequences is the same. The known structure is called the template, the unknown structure is called the target.

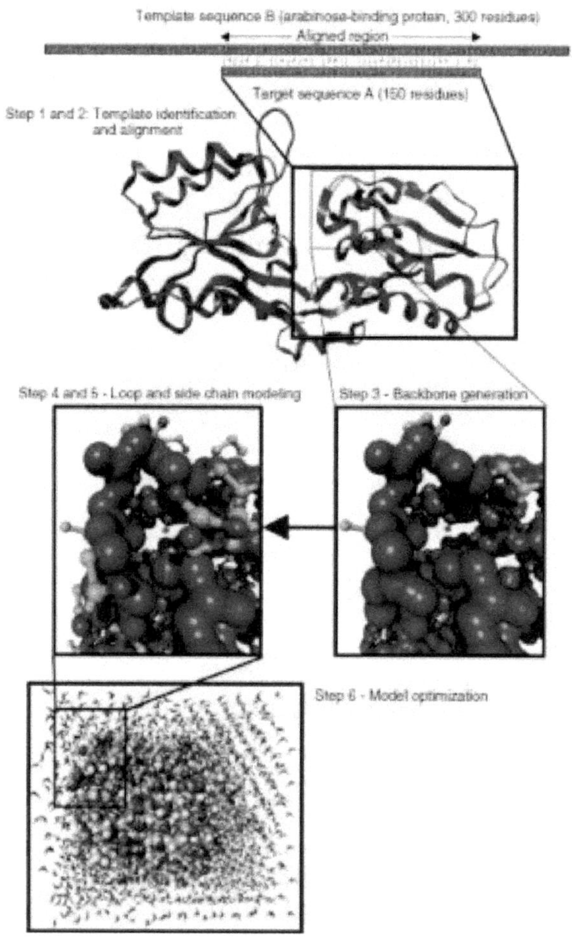

FUNCTION OF PROTEIN

- Proteins are complex molecules made up of chains of amino acids, and they are essential to the structure, function, and regulation of the body's cells, tissues, and organs. The function of proteins can be categorized into various roles:
- **Enzymatic Catalysis**: Proteins, in the form of enzymes, act as catalysts in biochemical reactions. Enzymes speed up chemical reactions without being consumed in the process. For example, **amylase** breaks down starch into sugars in the digestive system.
- **Structural Support**: Proteins provide structural integrity to cells and tissues. For example, **collagen** gives strength and flexibility to skin, tendons, and bones, while **keratin** is a major structural protein in hair and nails.
- **Transport and Storage**: Some proteins transport molecules throughout the body or store them. **Hemoglobin**, for instance, transports oxygen in the blood, while **ferritin** stores iron in cells.
- **Signaling and Regulation**: Proteins play a crucial role in cellular communication and signal transduction. For example, **hormones** like insulin regulate metabolism, and **receptors** on cell surfaces allow cells to respond to external signals.
- **Defense and Immunity**: Proteins are key components of the immune system. **Antibodies** (immunoglobulins) identify and neutralize foreign invaders like bacteria and viruses. **Complement proteins** also help to destroy pathogens.
- **Movement**: Proteins are responsible for cell movement and muscle contraction. **Actin** and **myosin** are the major proteins involved in muscle contraction, while **tubulin** forms microtubules that help in cell division and intracellular transport.
- **Gene Expression and Regulation**: Some proteins are involved in regulating gene expression by binding to DNA and influencing transcription. Examples include **transcription factors** that help initiate the process of gene transcription.
- **Cellular Homeostasis**: Proteins like **heat shock proteins** help in maintaining cellular homeostasis under stress conditions by assisting in the proper folding of other proteins and refolding denatured proteins.

Bioinformatics and Its Role in Protein Function

Bioinformatics is a field that combines biology, computer science, and information technology to understand biological data. It plays a significant role in studying proteins by analyzing their sequences, structures, functions, and interactions. Some of the key areas in bioinformatics related to protein study include:

- **Protein Sequence Analysis:** Bioinformatics tools are used to analyze protein sequences and predict their structure and function. Databases like **UniProt** and **NCBI Protein** provide information on protein sequences, annotations, and functional domains. Tools like **BLAST** (Basic Local Alignment Search Tool) allow researchers to compare protein sequences with known sequences in databases to infer similarities, evolutionary relationships, and possible functions.
- **Protein Structure Prediction:** Bioinformatics aids in predicting the 3D structure of proteins from their amino acid sequences. Algorithms like **AlphaFold** have made significant advancements in accurately predicting protein folding, which is crucial for understanding protein function and interaction.
- **Protein-Protein Interaction (PPI) Networks:** Understanding how proteins interact with one another is critical to comprehending cellular functions and pathways. Bioinformatics tools like **STRING** and **BioGRID** provide information about known PPIs and help predict interactions based on sequence similarity and experimental data. Protein interaction networks can be visualized using platforms like **Cytoscape**, helping to study how proteins function in concert within complex cellular systems.
- **Functional Annotation of Proteins:** Once a new protein sequence is discovered, bioinformatics tools are used to predict its potential function. This may involve identifying functional domains or motifs (specific sequences that correlate with a particular function), as well as predicting its cellular localization or involvement in specific pathways.
- **Proteomics Data Analysis:** In proteomics, which involves the large-scale study of proteins, bioinformatics tools help analyze mass spectrometry data to identify and quantify proteins in a sample. This includes identifying post-translational modifications (PTMs), which can alter protein function and activity.
- **Phylogenetic Analysis:** By comparing protein sequences across different species, bioinformatics can help track the evolutionary history of proteins. This information can provide insights into the conservation of protein functions across species.

- **Molecular Dynamics Simulations:** Bioinformatics is also used in modeling protein dynamics, simulating how proteins fold and interact in a dynamic environment. Software like **GROMACS** and **AMBER** are used to simulate molecular behavior in 3D, which can predict how proteins respond to environmental changes.
- **Systems Biology**: Bioinformatics integrates data from various biological experiments (genomics, transcriptomics, proteomics) to model entire biological systems. In this context, proteins are analyzed as part of larger networks, helping to understand their roles in cellular processes.

Applications of Bioinformatics in Protein Research

- **Drug Discovery**: Bioinformatics is instrumental in the design of drugs targeting specific proteins. By understanding the structure and function of a protein, researchers can design small molecules or biologics (e.g., monoclonal antibodies) that can modulate its activity, useful in treating diseases like cancer, diabetes, and viral infections.
- **Personalized Medicine**: By analyzing the proteins expressed in an individual's cells, bioinformatics can help tailor medical treatments to a person's genetic makeup and protein profile, which can lead to more effective therapies with fewer side effects.
- **Disease Biomarker Discovery**: Bioinformatics tools help identify proteins that are overexpressed or mutated in disease states, serving as potential biomarkers for diagnosis or prognosis. For instance, the identification of **BRCA1** and **BRCA2** proteins has been critical in assessing breast cancer risk.
- **Functional Genomics**: By studying proteins in the context of the genes that encode them, bioinformatics provides insights into gene function and regulation. This can uncover new biological pathways, offer insights into disease mechanisms, and lead to novel therapeutic strategies.

Proteins are crucial for a wide range of biological functions, and bioinformatics is an essential tool for understanding their structure, function, and interactions. Through computational analysis, researchers can predict how proteins work, interact, and contribute to disease, which can have profound implications for drug discovery, personalized medicine, and a deeper understanding of cellular biology.

12 Chapter
SYSTEM BIOLOGY AND PATHWAY ANALYSIS

INTRODUCTION

System biology and pathway analysis are integral fields of study that bridge the gap between molecular biology and computational science, offering profound insights into the complex networks that govern cellular processes. By integrating experimental data, computational models, and bioinformatics tools, systems biology seeks to understand the intricate interactions within biological systems at multiple levels, from genes and proteins to metabolic and signaling pathways. Pathway analysis, as a key component of this field, involves the systematic mapping and interpretation of molecular pathways, which are critical for regulating various cellular functions such as growth, metabolism, differentiation, and apoptosis. These pathways represent interconnected biochemical reactions, gene regulatory networks, and signal transduction cascades that maintain cellular homeostasis and enable responses to environmental changes or stresses. The ability to model and analyze these pathways in silico has revolutionized our approach to disease mechanisms, drug discovery, and personalized medicine. By elucidating the dynamic relationships between molecules, systems biology and pathway analysis allow for a more holistic understanding of biological systems, paving the way for novel therapeutic strategies and biomarkers for disease diagnosis and treatment.

BIOLOGICAL PATHWAYS AND NETWORKS

Biological pathways are interconnected series of biochemical reactions or molecular interactions that occur within a cell, leading to the production of a particular cellular outcome, such as the synthesis of molecules, energy production, or the regulation of gene expression. These pathways represent the cellular mechanisms that govern essential biological processes, from metabolism to signal transduction and gene regulation. The study of biological pathways has become a cornerstone of modern biology, particularly in fields like systems biology, bioinformatics, and molecular medicine. Biological networks, on the other hand, are abstract representations of the relationships and interactions between different biological entities such as genes, proteins, metabolites, or other molecules. These interactions, which may be biochemical, physical, or regulatory in nature, form intricate networks that govern cellular behavior. Biological networks help us understand how molecular interactions give rise to larger-scale processes like cell growth, differentiation, immune response, and disease development.

Types of Biological Pathways

Metabolic Pathways: These pathways consist of a series of chemical reactions that convert substrates into products, usually facilitated by enzymes. They include pathways such as glycolysis, the citric acid cycle (Krebs cycle), and the oxidative phosphorylation pathway. Metabolic pathways are essential for cellular energy production, synthesis of macromolecules, and maintenance of cell function.

- Example: Glycolysis is the breakdown of glucose into pyruvate, which is central to energy production in both aerobic and anaerobic conditions. The regulation of glycolysis is tightly controlled to meet the cell's energy demands.

Signal Transduction Pathways: These pathways involve the transmission of signals from the extracellular environment to the cell's interior, often involving receptors, secondary messengers, and protein kinases. Signal transduction is critical for coordinating responses to hormones, growth factors, environmental stress, and cellular communication.

- Example: The MAPK/ERK pathway is a signaling cascade that transmits signals from cell surface receptors to the nucleus, regulating cell growth, differentiation, and survival. Mutations in components of this pathway are frequently implicated in cancer.

Gene Regulatory Pathways: These pathways govern the expression of genes, ensuring that genes are turned on or off at the right time, in the right cell type, and in response to signals. Transcription factors, co-regulators, and chromatin modifiers are key players in regulating gene expression.

- Example: The NF-κB pathway plays a crucial role in immune response, inflammation, and cell survival. Dysregulation of this pathway can contribute to autoimmune diseases and cancer.

Protein-Protein Interaction Networks: These networks are formed by the physical interactions between proteins that enable cellular processes such as signal transduction, metabolism, and gene expression. The study of protein-protein interactions (PPIs) is fundamental in understanding cellular function.

- Example: In the p53 pathway, the tumor suppressor protein p53 interacts with a variety of proteins to regulate cell cycle arrest, DNA repair, and apoptosis in response to cellular stress. Its malfunction often leads to uncontrolled cell proliferation and cancer.

Post-Translational Modification Pathways: These pathways involve modifications to proteins after they are synthesized, affecting their function, localization, and interactions. Modifications include phosphorylation, acetylation, methylation, and ubiquitination.

- Example: Ubiquitin-proteasome pathway is responsible for tagging proteins for degradation. This pathway regulates various cellular processes, including the cell cycle and protein quality control.

Components of Biological Networks

- **Nodes:** Represent individual biological entities, such as genes, proteins, metabolites, or even entire cells. In gene regulatory networks, the nodes may represent specific genes or transcription factors. In protein interaction networks, the nodes represent individual proteins.
- **Edges:** The connections between nodes, which represent interactions or relationships. For instance, in metabolic networks, edges might represent chemical reactions, while in signaling networks, edges may indicate the activation or inhibition between proteins or other molecules.
- **Cyclic and Non-Cyclic Networks:** Biological networks can be cyclic (such as feedback loops in signal transduction or metabolic cycles like the Krebs cycle) or non-cyclic (such as linear signaling pathways or gene expression networks).

Types of Biological Networks

1. **Gene Regulatory Networks:** These are networks of genes that regulate one another's expression through transcription factors, co-regulators, and signaling molecules. Gene regulatory networks control cellular processes such as development, differentiation, and response to environmental cues.
2. **Protein-Protein Interaction (PPI) Networks:** PPI networks map the physical interactions between proteins, essential for cellular processes such as signal transduction, molecular transport, and cellular organization. These networks help us understand how proteins collaborate to carry out complex cellular functions.
3. **Metabolic Networks:** These are networks of metabolic reactions that interconnect enzymes, substrates, and products. The study of metabolic networks allows for understanding how cells produce energy, synthesize molecules, and maintain homeostasis.
4. **Signaling Networks:** These are networks of molecular signals that orchestrate responses to environmental and intracellular cues. Signaling networks involve the activation and inhibition of receptors, secondary messengers, kinases, and other signaling molecules, often culminating in changes in gene expression and cell behavior.

5. **Disease Networks:** These networks are often constructed to identify key molecular drivers of diseases like cancer, diabetes, or neurodegenerative disorders. Disease networks connect mutations, altered protein functions, and disrupted pathways to specific pathological states.

Pathway and Network Analysis Tools

Several computational tools have been developed to facilitate the analysis of biological pathways and networks. These tools help identify key nodes and edges, predict the effects of perturbations, and discover potential biomarkers or therapeutic targets.

- **KEGG (Kyoto Encyclopedia of Genes and Genomes):** A comprehensive database that provides information on metabolic pathways, signal transduction pathways, and other biological networks.
- **Reactome:** A curated database of biological pathways that covers a wide array of human cellular processes.
- **Cytoscape:** A popular open-source software for visualizing and analyzing molecular interaction networks, including gene and protein interactions, metabolic networks, and more.
- **Gene Ontology (GO):** A framework for the classification of gene functions, often used in pathway and network analysis to interpret gene expression data in the context of biological processes.
- **Ingenuity Pathway Analysis (IPA):** A commercial tool that helps researchers analyze data sets within the context of known biological pathways, enabling hypothesis generation in the study of disease mechanisms.

Importance of Biological Pathways and Networks

- **Understanding Cellular Mechanisms:** Pathway and network analysis provides a comprehensive view of how cells maintain their functions, respond to external stimuli, and adapt to changing environments.
- **Disease Mechanisms:** By mapping the molecular interactions underlying diseases, researchers can identify key regulatory nodes that may serve as therapeutic targets. For example, understanding the disruption of signaling pathways in cancer has led to targeted therapies like kinase inhibitors.
- **Drug Discovery and Personalized Medicine:** Pathway analysis is increasingly used to identify drug targets and biomarkers. By mapping a patient's molecular profile to known disease pathways, personalized treatment strategies can be developed, improving therapeutic outcomes.
- **Functional Genomics:** Understanding how genes interact within a network can shed light on their functions and how mutations in specific genes

contribute to disease. This is particularly valuable in studying complex diseases like cancer, cardiovascular disease, and neurological disorders.

Biological pathways and networks form the backbone of cellular function, orchestrating the dynamic and highly integrated processes that govern life. The ability to study these pathways at a systems level enables a deeper understanding of biology, reveals the molecular underpinnings of diseases, and opens up new avenues for drug discovery, diagnostics, and personalized medicine. As tools and technologies evolve, pathway and network analysis will continue to play a pivotal role in advancing both basic biological research and clinical applications.

METABOLIC PATHWAYS ANALYSIS

Metabolic pathways are intricate networks of biochemical reactions that occur within living cells to maintain life processes, including energy production, biosynthesis, and the regulation of cell functions. These pathways are composed of a series of interconnected enzyme-catalyzed reactions that convert substrates (often metabolites) into products. Each pathway is tightly regulated to ensure that cellular processes operate efficiently and respond dynamically to both internal and external signals. The study and analysis of metabolic pathways is crucial for understanding fundamental biological processes, cellular homeostasis, and disease mechanisms. Metabolic pathway analysis, a key component of systems biology, integrates experimental data with computational models to map and interpret these complex networks. By examining the flow of metabolites and the activity of enzymes within these pathways, researchers can gain insights into metabolic dysregulation, identify biomarkers for disease, and develop therapeutic strategies targeting specific metabolic processes.

Concepts in Metabolic Pathway Analysis

- **Metabolites:** Metabolites are small molecules that are either substrates or products in biochemical reactions. They include a wide variety of compounds such as amino acids, nucleotides, lipids, carbohydrates, and cofactors.
- **Enzymes:** Enzymes are proteins that catalyze biochemical reactions, speeding up the conversion of substrates into products. In metabolic pathways, enzymes ensure that reactions occur at appropriate rates and under appropriate conditions.
- **Pathway Intermediates:** These are molecules that are produced as part of a pathway but are not the final products. They play essential roles in driving subsequent reactions, and their accumulation or depletion can indicate disruptions in pathway activity.

- **Energy Carriers:** Many metabolic pathways are involved in energy production or consumption. Key molecules like **ATP** (adenosine triphosphate), **NADH** (nicotinamide adenine dinucleotide), and **FADH2** are central to the transfer of energy within the cell.
- **Regulatory Mechanisms:** Metabolic pathways are regulated by enzymes that control the speed and direction of biochemical reactions. These regulatory mechanisms include feedback inhibition, allosteric regulation, and covalent modifications, which allow cells to adapt to changing environmental conditions or nutrient availability.

Major Types of Metabolic Pathways

Catabolic Pathways (Energy-Producing)

These pathways break down larger molecules into smaller ones, releasing energy that can be captured by the cell. Catabolic pathways are crucial for energy production, especially in processes like **aerobic respiration** and **fermentation.**

Example:

- **Glycolysis:** This is the breakdown of glucose into pyruvate, producing small amounts of ATP. It occurs in the cytoplasm and is the first step in both aerobic and anaerobic respiration.
- **Citric Acid Cycle (Krebs Cycle):** This cycle, occurring in the mitochondria, further processes pyruvate into carbon dioxide and high-energy electron carriers (NADH, FADH2).
- **Oxidative Phosphorylation:** This pathway, also in the mitochondria, generates the majority of ATP in aerobic organisms through the electron transport chain and chemiosmosis.

Anabolic Pathways (Biosynthetic)

Anabolic pathways use energy to build complex molecules from simpler precursors. These pathways are crucial for cell growth, maintenance, and repair. Anabolism requires input of energy, usually in the form of ATP, NADPH, or other high-energy molecules.

Example:

- **Gluconeogenesis:** The process by which glucose is synthesized from non-carbohydrate precursors like lactate or amino acids, often occurring in the liver.
- **Protein Synthesis:** The assembly of amino acids into proteins, which involves metabolic pathways like **translation** and **amino acid metabolism.**

Intermediary Metabolism

This refers to the network of interconnected pathways involved in the synthesis and breakdown of metabolites, serving as intermediates between catabolism and anabolism. It includes reactions involved in the metabolism of amino acids, lipids, and carbohydrates.

Example

- **Pentose Phosphate Pathway:** An alternative pathway for glucose metabolism that generates nucleotides and NADPH, which are essential for DNA synthesis and antioxidant defense.

Detoxification and Waste Removal

Many metabolic pathways are involved in the removal of toxic metabolites or the detoxification of harmful substances.

Example

- **Urea Cycle:** This pathway converts ammonia (a toxic byproduct of amino acid metabolism) into urea, which is then excreted from the body.

Metabolic Pathway Analysis Methods

Metabolic pathway analysis involves several approaches to studying the structure, function, and regulation of metabolic networks. Some common methods include:

Metabolomics

Metabolomics is the study of the complete set of metabolites in a biological sample, providing a snapshot of metabolic activity. Advanced techniques like **mass spectrometry (MS)** and **nuclear magnetic resonance (NMR)** spectroscopy are used to profile metabolites and their concentrations.

Applications:

- Identifying metabolic changes in diseases (e.g., cancer metabolism, diabetes).
- Studying the effects of drugs or environmental changes on metabolism.
- Discovering biomarkers for disease diagnosis or prognosis.

Flux Analysis

Flux analysis focuses on measuring the rates at which metabolites flow through metabolic pathways. This method involves isotopic labeling, where specific atoms (e.g., carbon-13) are incorporated into metabolites, allowing researchers to trace their movement through the pathway.

Application:

- Understanding how cells allocate resources under different conditions (e.g., energy production vs. biosynthesis).
- Investigating how metabolic fluxes change in response to nutrient availability or disease states.

Pathway Database Integration

Several databases catalog known metabolic pathways, offering a framework for analyzing experimental data. These include:

- **KEGG (Kyoto Encyclopedia of Genes and Genomes):** A comprehensive resource for metabolic pathways, enzyme functions, and molecular networks.
- **Reactome:** A database that includes curated pathways, including many related to metabolism, and allows for pathway enrichment analysis.
- **MetaCyc:** A database specifically focused on metabolic pathways, providing detailed descriptions of metabolic reactions across different organisms.

Computational Modeling and Simulation

Metabolic network models are mathematical representations of metabolic pathways, often used to predict how cells will behave under different conditions. These models can be built using data from genome sequencing, transcriptomics, proteomics, and metabolomics.

Methods:

- **Stoichiometric models:** These models describe the relationship between metabolites and reactions using a system of linear equations.
- **Constraint-based optimization:** This technique uses linear programming to find the optimal distribution of fluxes through the network that satisfies certain constraints (e.g., maximizing ATP production).
- **Dynamic models:** These models simulate changes in metabolite concentrations over time, incorporating kinetic parameters for enzyme reactions.

Gene Knockdown/Overexpression Studies

Altering the expression of genes encoding enzymes involved in specific metabolic pathways can help reveal their roles in metabolism. Techniques like RNA interference (RNAi) or CRISPR-Cas9 can be used to knock down or knock out genes, allowing for the study of metabolic changes in response to these genetic modifications.

Pathway Enrichment Analysis

This approach is used to identify biological pathways that are significantly altered in response to experimental conditions (e.g., disease states, treatments). Pathway enrichment analysis often involves comparing experimental data (e.g., from transcriptomics or proteomics) against predefined pathway databases to identify overrepresented metabolic pathways.

Applications of Metabolic Pathway Analysis

- **Cancer Metabolism:** Cancer cells often exhibit altered metabolic pathways to support rapid growth, including increased glycolysis (known as the **Warburg effect**), altered lipid metabolism, and enhanced nucleotide biosynthesis. Metabolic pathway analysis helps in identifying potential targets for cancer therapy.
- **Diabetes:** In diabetes, particularly type 2, disruptions in glucose and lipid metabolism lead to insulin resistance and other complications. Metabolic pathway analysis can help identify biomarkers for early detection and monitor therapeutic interventions.
- **Obesity and Metabolic Syndrome:** Disruptions in metabolic pathways, including impaired lipid metabolism and mitochondrial dysfunction, are central to obesity and metabolic syndrome. Understanding these pathways can lead to better treatments or prevention strategies.
- **Neurodegenerative Diseases:** Alterations in metabolic pathways are implicated in neurodegenerative diseases such as Alzheimer's and Parkinson's. Metabolic pathway analysis may identify metabolic biomarkers for early diagnosis or potential targets for intervention.
- **Drug Discovery and Toxicology:** Drug-induced metabolic changes can have profound effects on cellular function. Metabolic profiling can identify potential drug side effects and guide the development of safer and more effective therapies.

Metabolic pathway analysis plays a pivotal role in understanding the molecular mechanisms that sustain life, as well as in identifying metabolic dysregulation associated with diseases. By integrating experimental data with computational models, researchers can uncover the complex networks of biochemical reactions that drive cellular processes and discover novel strategies for treating metabolic disorders, cancers, and other diseases. With advances in technologies like metabolomics, flux analysis, and computational modeling, the ability to analyze and manipulate metabolic pathways continues to evolve, offering new opportunities for precision medicine and therapeutic intervention.

INTEGRATION OF OMICS DATA

The integration of omics data refers to the process of combining and analyzing data from multiple "omics" layers of biological information—such as genomics, transcriptomics, proteomics, metabolomics, and others—to gain a comprehensive understanding of biological systems. Each of these omics fields provides a distinct perspective on biological function and regulation. However, when these data sets are integrated, they offer a more holistic view of cellular processes, disease mechanisms, and organismal biology, allowing researchers to identify interactions across different biological scales. The advent of high-throughput technologies has made it possible to generate large amounts of data from different omics layers, but these datasets are often complex, heterogeneous, and voluminous. Thus, integrating omics data requires both computational and statistical tools to combine these diverse data sources in meaningful ways. The goal of omics data integration is to uncover relationships between genes, proteins, metabolites, and phenotypic traits, which would otherwise remain hidden when considering each layer in isolation.

Omics Layers and Their Roles

Genomics

Genomics refers to the study of the complete set of genes or genomes within an organism. It involves sequencing the DNA to identify genetic variations, mutations, and structural alterations that may affect an organism's traits or predisposition to diseases.

- **Techniques:** Next-generation sequencing (NGS), whole-genome sequencing (WGS), and genome-wide association studies (GWAS).
- **Applications:** Identifying genetic risk factors for diseases, understanding genetic diversity, and mapping genetic pathways involved in health and disease.

Transcriptomics

Transcriptomics is the study of the transcriptome, which consists of all the RNA molecules produced in a cell. It measures gene expression levels by quantifying messenger RNA (mRNA), revealing how genes are regulated in different tissues or under varying conditions.

- **Techniques:** RNA sequencing (RNA-seq), microarrays, and quantitative PCR.
- **Applications:** Identifying genes involved in disease pathways, assessing gene expression profiles in different conditions, and understanding cellular responses to external stimuli.

Proteomics

Proteomics focuses on the study of the proteome, the entire set of proteins expressed by a genome at a given time. Since proteins carry out the majority of cellular functions, proteomics provides insights into protein abundance, post-translational modifications (PTMs), protein-protein interactions, and cellular localization.

- **Techniques:** Mass spectrometry (MS), 2D gel electrophoresis, and protein microarrays.
- **Applications:** Identifying biomarkers for disease, studying protein function and interaction networks, and exploring cellular processes like signal transduction and metabolism.

Metabolomics

Metabolomics is the study of the metabolome, which consists of all the small molecules (metabolites) in a biological system, including sugars, lipids, amino acids, and hormones. These metabolites are the products of cellular processes and reflect the functional state of the cell or organism.

- **Techniques:** Nuclear magnetic resonance (NMR) spectroscopy, mass spectrometry (MS), and chromatography.
- **Applications:** Identifying metabolic biomarkers, understanding disease-related metabolic changes, and exploring cellular energy status or nutrient utilization.

Lipidomics

Lipidomics is the study of the lipidome, which includes all lipids in a cell or organism. Lipids are crucial for cellular membrane structure, energy storage, and signaling.

- **Techniques:** Mass spectrometry, liquid chromatography (LC-MS), and gas chromatography.
- **Applications:** Understanding lipid metabolism in diseases such as obesity, diabetes, and neurodegenerative disorders, and exploring lipid-based signaling mechanisms.

Epigenomics

Epigenomics involves the study of epigenetic modifications, such as DNA methylation, histone modification, and non-coding RNAs, which can influence gene expression without altering the underlying genetic code.

- **Techniques:** ChIP-sequencing, bisulfite sequencing, and methylation arrays.

- **Applications:** Studying gene regulation, cellular differentiation, and disease mechanisms like cancer and developmental disorders.

Challenges in Omics Data Integration

The integration of omics data from different biological layers is a highly complex process due to several inherent challenges:

- **Heterogeneity of Data:** Data from different omics technologies have different formats, scales, and types of information. For instance, genomics data might be in the form of sequences, transcriptomics data as expression levels, and proteomics data as quantitative protein abundances. Integrating such diverse types of data requires specialized tools that can harmonize and relate these different data types.
- **Dimensionality and Sparsity:** Omics data sets are often high-dimensional, containing tens of thousands of features (genes, proteins, metabolites), but only a relatively small number of samples. This "curse of dimensionality" can make it difficult to find meaningful patterns without overfitting or losing generalizability. Additionally, the data is often sparse, with missing values due to technical limitations or experimental design.
- **Temporal and Spatial Variability:** Many omics datasets come from experiments conducted under specific conditions or at particular time points. Integrating data across different experimental setups, time points, or tissues requires careful consideration of these variables to avoid biases.
- **Data Quality and Preprocessing:** Omics datasets often contain noise, technical errors, or batch effects. Preprocessing steps such as normalization, quality control, and imputation of missing data are essential before integrating data from different sources to ensure meaningful and reliable results.
- **Complex Interactions:** Biological systems are highly complex, and the relationships between genes, proteins, metabolites, and their functions can be nonlinear, hierarchical, and context-dependent. Developing computational models that can accurately capture these interactions is a major challenge.

Approaches to Omics Data Integration

Several computational and statistical methods have been developed to overcome these challenges and integrate omics data:

- **Multivariate Analysis:** Multivariate techniques such as **principal component analysis (PCA)**, **canonical correlation analysis (CCA)**, and **partial least squares (PLS)** are commonly used to reduce dimensionality and explore patterns across multiple omics layers. These methods help identify latent variables or common patterns across different data sets.

- **Network-Based Approaches**: Network-based approaches construct interaction networks (e.g., protein-protein interaction networks or gene regulatory networks) to link genes, proteins, and metabolites across different layers. By integrating omics data with these networks, researchers can explore the functional relationships between various biological components. **Example**: Integrating gene expression data with protein interaction networks can help identify key proteins that are central to specific pathways or diseases.
- **Machine Learning:** Machine learning (ML) techniques, including supervised and unsupervised methods, are increasingly applied to integrate omics data. Methods like **random forests**, **support vector machines (SVM)**, and **deep learning** can help model complex interactions, predict outcomes, and classify disease states by integrating multiple omics layers.
- **Pathway and Gene Set Enrichment Analysis**: Tools like **Gene Set Enrichment Analysis (GSEA)** and **Ingenuity Pathway Analysis (IPA)** help integrate omics data by mapping genes or proteins onto biological pathways. These methods can identify which biological processes or pathways are most significantly affected under different conditions.
- **Integration of Time-Series Data**: Integrating omics data across different time points requires methods that can account for temporal dynamics. **Dynamic modeling** approaches, such as **ordinary differential equations (ODE)** and **Bayesian networks**, are often used to model the time-dependent behavior of biological systems and integrate data from multiple time points.
- **Data Fusion Approaches:** Data fusion integrates omics data from multiple sources into a unified model. This could include a combination of multi-omics layers (e.g., genomics, transcriptomics, proteomics) or data from different experimental platforms (e.g., RNA-seq, microarrays, mass spectrometry).

Applications of Omics Data Integration

- **Disease Mechanism Understanding**: Integrating omics data helps reveal the molecular underpinnings of diseases like cancer, neurodegenerative disorders, and metabolic diseases. For instance, combining transcriptomics with proteomics and metabolomics can show how gene expression changes are linked to protein activity and metabolic shifts in disease states.
- **Biomarker Discovery**: Multi-omics integration enables the identification of disease-specific biomarkers that might not be evident when studying

a single omics layer. These biomarkers can be used for early diagnosis, prognosis, and personalized treatment.

- **Drug Discovery and Development:** By integrating omics data, researchers can identify potential drug targets, predict drug responses, and assess off-target effects. For example, integrating proteomics with metabolomics could help understand how a drug affects cellular metabolism at the molecular level.
- **Personalized Medicine:** Omics data integration is essential for precision medicine, as it enables the development of tailored therapeutic strategies based on a patient's unique molecular profile. For instance, integrating genomics, transcriptomics, and proteomics can predict how an individual will respond to certain treatments.
- **System Biology and Modeling:** Integrating multiple omics layers is fundamental to building accurate systems biology models. These models can predict cellular behavior under different conditions and simulate responses to genetic, environmental, or therapeutic perturbations.

NEXT-GENERATION SEQUENCING (NSG) TECHNOLOGIES

Next-Generation Sequencing (NGS) technologies represent a transformative leap in genomic research, enabling the rapid, high-throughput sequencing of DNA and RNA at an unprecedented scale and accuracy. Unlike traditional Sanger sequencing, which is time-consuming and limited in throughput, NGS technologies allow for the simultaneous sequencing of millions of DNA fragments, generating vast amounts of data in a fraction of the time. This has revolutionized fields such as genomics, personalized medicine, disease research, and drug discovery by providing deeper insights into genetic variations, gene expression profiles, and complex biological processes. NGS platforms utilize parallel sequencing techniques, often relying on technologies like sequencing by synthesis, ion semiconductor sequencing, or single-molecule sequencing, each offering unique advantages in terms of read length, error rate, and application versatility. The affordability, speed, and scalability of NGS have made it a cornerstone of modern molecular biology, facilitating everything from whole-genome sequencing to transcriptome analysis and beyond. With continuous advancements in NGS technologies, the ability to explore the intricate details of the genome is becoming increasingly accessible, opening new frontiers in understanding human health, evolution, and disease mechanisms.

NSG PLATFORMS: AN OVERVIEW

Next-Generation Sequencing (NGS) technologies have revolutionized genomics by providing high-throughput, scalable, and cost-effective solutions for sequencing

DNA and RNA. These platforms enable the sequencing of entire genomes, exomes, transcriptomes, and more, delivering vast amounts of data quickly and efficiently. The variety of NGS platforms available today offer distinct strengths, such as read length, throughput, accuracy, and cost, making them suitable for different applications across research, clinical diagnostics, and personalized medicine. NGS platforms can be broadly categorized based on their sequencing technology and the methods used to generate data. Some of the most commonly used NGS platforms include **Illumina**, **Ion Torrent**, **Pacific Biosciences (PacBio)**, **Oxford Nanopore**, and **BGI's MGISEQ** systems. Each of these platforms uses a different sequencing approach and has its own set of advantages and limitations.

Major NGS Platforms

Illumina Sequencing Platforms

Illumina is by far the most widely used NGS technology, known for its sequencing by synthesis (SBS) method. It provides high-throughput sequencing with high accuracy, making it ideal for a wide range of applications, including whole-genome sequencing, targeted sequencing, and RNA sequencing.

- **Technology**: Sequencing by Synthesis (SBS) Illumina platforms use a technique where DNA is fragmented, and adapters are added to the fragments. These are then immobilized on a solid surface (flow cell), and nucleotides are incorporated one by one during sequencing. Fluorescently labeled nucleotides are used to detect incorporation events in real time, allowing for highly accurate base calling.
- **Key Features**:
  - High throughput (millions to billions of reads in a single run)
  - Short to medium read lengths (75 to 300 base pairs)
  - High accuracy and low error rates
  - Versatile applications, including whole-genome sequencing, targeted sequencing, RNA-seq, and metagenomics.
  - Widely used in both clinical and research settings
- **Common Illumina Platforms**:
  - **MiSeq**: Ideal for small to medium sequencing projects, such as amplicon sequencing, 16S rRNA sequencing, and targeted sequencing.
  - **NextSeq**: Provides high-throughput sequencing for whole genome and exome sequencing, with read lengths of up to 150 bp.

- **NovaSeq:** Offers ultra-high throughput sequencing for large-scale genomic studies, such as large cohort studies or human genome sequencing.

Ion Torrent (Thermo Fisher Scientific)

Ion Torrent sequencing uses a different technology from Illumina's SBS approach, based on semiconductor sequencing. It measures the release of protons as nucleotides are incorporated into the growing DNA strand. This technology is known for its speed and simplicity, with the ability to sequence in real-time.

- **Technology:** Ion Semiconductor Sequencing In Ion Torrent sequencing, the DNA is fragmented and attached to beads, which are then amplified. During sequencing, as each nucleotide is added to the growing strand, it causes a proton release that is detected by a semiconductor chip, allowing the system to identify the incorporated base.
- **Key Features:**
  - Real-time sequencing, faster than other platforms
  - Read lengths range from 200 to 400 base pairs
  - Smaller, more affordable instruments, making it suitable for smaller labs and point-of-care sequencing
  - Good for targeted sequencing and sequencing of smaller genomes
  - Often used in clinical diagnostics and microbial sequencing
- **Common Ion Torrent Platforms:**
  - **Ion Proton:** Offers high-throughput sequencing for large genomes, such as human genome sequencing.
  - **Ion S5:** A smaller, more affordable platform for medium-throughput sequencing.

PacBio (Pacific Biosciences)

Pacific Biosciences (PacBio) is known for its **Single Molecule Real-Time (SMRT)** sequencing technology, which can produce very long read lengths and is particularly useful for sequencing complex or repetitive regions in genomes. PacBio sequencing is also well-suited for applications such as full-length transcript sequencing, structural variant detection, and de novo genome assembly.

- **Technology:** Single Molecule Real-Time (SMRT) Sequencing PacBio uses a method where individual DNA molecules are sequenced in real time as they pass through a nanopore-like hole in a specialized chip. The system detects fluorescent signals emitted by each base as it is incorporated into the growing DNA strand. The key advantage is the ability to generate ultra-long reads (up to tens of kilobases).

- **Key Features:**
  - Long read lengths (up to 50,000 bp) enable the resolution of complex regions, structural variants, and repetitive sequences.
  - High accuracy for base calling in the presence of long repetitive sequences.
  - More expensive than other platforms, but highly valuable for specific applications such as de novo assembly and metagenomics.
  - Suitable for applications that require sequencing through challenging genomic regions (e.g., highly repetitive or GC-rich regions).
- **Common PacBio Platforms:**
  - **Sequel II System:** Provides long-read sequencing and improved throughput compared to earlier PacBio systems.
  - **RS II System:** Older platform for long-read sequencing but still used in specific applications.

Oxford Nanopore Technologies

Oxford Nanopore sequencing is another long-read platform that uses a radically different approach. It sequences DNA or RNA by passing it through a nanopore and measuring changes in electrical current as each nucleotide passes through. This technology is particularly notable for its portability and real-time sequencing capabilities.

- **Technology:** Nanopore Sequencing.0In Oxford Nanopore sequencing, DNA or RNA is fed through a tiny nanopore embedded in a membrane. As each nucleotide passes through, it generates a distinct electrical signal. These signals are analyzed to identify the nucleotide sequence.
- **Key Features:**
  - Ultra-long reads (up to 2 million bases or more) that can span entire chromosomes or even full-length transcripts.
  - Real-time sequencing with the ability to analyze data as it is being generated.
  - Portability (e.g., the **MinION** device is a small, USB-powered sequencer that can be used in the field).
  - Can sequence both DNA and RNA, enabling a broad range of applications in genomics, transcriptomics, and metagenomics.

- Lower throughput compared to Illumina, but growing in popularity for specific use cases, such as field-based sequencing and rapid pathogen identification.

◉ **Common Oxford Nanopore Platforms:**

- **MinION:** Portable, USB-powered device designed for field applications and small sequencing projects.
- **GridION:** A benchtop device offering higher throughput than MinION for larger sequencing projects.
- **PromethION:** High-throughput sequencer designed for large-scale genomic studies.

MGISEQ (BGI)

BGI's **MGISEQ** platform is another high-throughput NGS technology that competes with Illumina. It uses a similar sequencing-by-synthesis approach but with some different optimizations in chemistry and instrumentation.

◉ **Technology:** Sequencing by Synthesis MGISEQ uses a technology based on sequencing by synthesis, similar to Illumina, but incorporates proprietary chemistry and optics to enhance performance and reduce costs.

◉ **Key Features:**

- Competitive throughput and accuracy similar to Illumina platforms.
- Cost-effective, with potential for use in large-scale sequencing projects.
- Offers versatility in applications such as whole-genome sequencing, RNA-seq, and exome sequencing.

◉ **Common MGISEQ Platforms:**

- **MGISEQ-2000:** High-throughput platform for genomic and transcriptomic applications.
- **MGISEQ-T7:** Designed for large-scale sequencing projects, providing high throughput with cost-efficiency.

Comparison of NGS Platforms

Feature	Illumina	Ion Torrent	PacBio	Oxford Nanopore	MGISEQ
Sequencing Technology	Sequencing by Synthesis (SBS)	Semiconductor Sequencing	Single Molecule Real-Time	Nanopore Sequencing	Sequencing by Synthesis
Read Length	Short to Medium (75-300 bp)	Short (200-400 bp)	Long (up to 50,000 bp)	Ultra Long (up to 2 Mbp)	Short to Medium (up to 150 bp)

Feature	Illumina	Ion Torrent	PacBio	Oxford Nanopore	MGISEQ
Accuracy	High	Moderate	High (with Circular Consensus Sequencing)	Moderate to High (depends on technology)	High
Throughput	Very High	Moderate	Moderate	Moderate to Low	High
Cost	Moderate	Moderate to Low	High	Low to Moderate	Moderate
Applications	Whole-genome, RNA-seq, Exome	Targeted sequencing, Microbial	De novo assembly, Long reads	Long reads, Real-time, RNA-seq	Whole-genome, RNA-seq

Each NGS platform offers unique strengths, making them suited to specific research needs and applications. Illumina remains the gold standard for high-throughput sequencing, PacBio and Oxford Nanopore stand out for their ability to generate long reads, and Ion Torrent provides fast, cost-effective sequencing. As technology continues to evolve, these platforms will continue to shape the

NGS DATA ANALYSIS PIPELINES

Next-Generation Sequencing (NGS) technologies generate vast amounts of high-dimensional data, which require sophisticated bioinformatics pipelines for analysis. These pipelines are designed to process, analyze, and interpret raw sequencing data, transforming it into meaningful biological insights. NGS data analysis involves multiple steps, from quality control and alignment to variant calling, annotation, and downstream biological interpretation. Each step in the pipeline is crucial for ensuring the accuracy, reproducibility, and comprehensiveness of the results. The complexity of NGS data analysis depends on the type of sequencing performed (e.g., whole-genome sequencing, RNA sequencing, or targeted sequencing) and the specific research question. Despite variations in application, most NGS analysis pipelines follow a general workflow. Below is a detailed description of the typical steps involved in NGS data analysis pipelines.

Steps in NGS Data Analysis Pipelines

Data Quality Control (QC) and Preprocessing

The first step in any NGS analysis is to assess the quality of the raw sequence data. Raw reads from NGS platforms are often noisy, contain adapter sequences, low-quality bases, or even sequencing artifacts. **Quality control (QC)** helps identify issues in the data that might affect the reliability of downstream analyses.

- **Quality Checking Tools:** Tools like **FastQC** or **MultiQC** are commonly used to evaluate read quality, base composition, GC content, and the presence of adapters or contamination.
- **Trimming and Filtering:** Low-quality reads and bases are removed using tools like **Trimmomatic** or **Cutadapt.** Additionally, adapter sequences and short, non-informative reads can be filtered out to reduce bias and improve the quality of subsequent analysis.

Read Alignment/Mapping

After quality control and preprocessing, the next step is to align the high-quality reads to a reference genome or transcriptome. **Read alignment** helps identify where each sequencing read originated in the genome and is crucial for variant discovery and quantifying gene expression.

- **Alignment Algorithms:** Algorithms like **Bowtie2, BWA (Burrows-Wheeler Aligner)**, or **STAR** (for RNA-seq) are commonly used for aligning reads to a reference genome.
- **Reference Genome:** The choice of reference genome is critical. A reference genome provides a baseline to which the sequenced reads are aligned. For example, human genome references like **hg19** or **hg38** are commonly used for human NGS data.
- **Alignment Types:** The two primary types of alignment are **short-read alignment** (e.g., for DNA-seq or ChIP-seq) and **spliced alignment** (e.g., for RNA-seq, to account for exons and introns).

Post-Alignment Processing

Once the reads are aligned, several additional steps are required to further refine the alignment and improve the accuracy of downstream analyses.

- **Sorting:** After alignment, reads are often sorted by their position in the genome to facilitate downstream analyses. Tools like **SAMtools** or **Picard** are used to sort alignments into **BAM** (Binary Alignment/Map) files.
- **Marking Duplicates:** PCR amplification during sequencing can lead to duplicate reads. These duplicates are usually marked using **Picard** tools to avoid inflating read counts during variant calling.
- **Indel Realignment:** Small insertions and deletions (indels) are common in NGS data and can introduce misalignments. Tools like **GATK (Genome Analysis Toolkit)** provide realignment steps to correct these indel-related errors.

- **Base Quality Score Recalibration:** This process involves adjusting the quality scores of base calls based on known systematic biases in the sequencing data. **GATK** provides a popular method for this recalibration.

Variant Calling

After alignment and processing, the next step is to identify genetic variants, such as single nucleotide polymorphisms (SNPs) and insertions/deletions (indels). Variant calling involves comparing the aligned reads against a reference genome to detect differences.

- **Tools for Variant Calling:** Commonly used tools for variant calling include:
  - **GATK HaplotypeCaller:** A widely used tool for calling SNPs and indels in whole-genome or exome sequencing data.
  - **Samtools:** A tool for working with BAM files, which also includes variant calling functionality.
  - **FreeBayes:** Another variant caller that works by using Bayesian models to detect SNPs and indels.
- **Types of Variants:** Variant calling identifies different types of genetic variations:
  - **SNPs:** Single nucleotide changes (e.g., A to T).
  - **Indels:** Insertions or deletions of nucleotides.
  - **Structural Variants:** Larger genomic alterations, such as deletions, duplications, or translocations (detected using specialized tools like **Manta** or **Delly**).

Variant Annotation

After calling variants, the next step is to annotate them to interpret their biological significance. Annotation provides information about the location of variants (e.g., whether they are located in coding regions, introns, or regulatory regions) and their potential impact on gene function.

- **Annotation Tools:**
  - **ANNOVAR** and **VEP (Variant Effect Predictor)** are widely used tools for annotating genetic variants with information such as gene names, functional effects (e.g., missense, silent mutations), and known associations with diseases.
  - **dbSNP** and **ClinVar:** Public databases containing information on common and clinically significant variants.

- **Impact Assessment:** The annotation step also involves assessing the potential functional impact of the variant, using algorithms like **SIFT**, **PolyPhen**, or **CADD**, which predict whether a mutation might affect protein function.

Gene Expression Analysis (For RNA-seq)

In RNA sequencing (RNA-seq), the goal is often to measure gene expression levels. Once the reads are aligned, the next step is quantification of gene expression.

- **Gene Expression Quantification Tools:** Tools like **HTSeq, Cufflinks**, and **featureCounts** are used to calculate the number of reads mapped to each gene or transcript, which can then be used to estimate gene expression levels.
- **Normalization:** To make gene expression data comparable across different samples, normalization methods such as **RPKM** (Reads Per Kilobase per Million mapped reads), **FPKM** (Fragments Per Kilobase of exon per Million fragments mapped), or **TPM** (Transcripts Per Million) are used to account for sequencing depth and gene length.
- **Differential Expression Analysis:** Once expression data is quantified, tools like **DESeq2**, **edgeR**, or **limma** can be used to identify genes that are differentially expressed between different conditions or groups, such as diseased vs. healthy samples.

Data Visualization

Data visualization is a critical part of NGS data analysis, as it helps researchers interpret and communicate their findings. Key visualization techniques include:

- **Genome Browser:** Tools like **IGV (Integrative Genomics Viewer)** or **UCSC Genome Browser** allow researchers to visualize aligned reads, variants, and gene annotations in the context of the genome.
- **Variant Visualization: Circos plots** or **Manhattan plots** are often used to visualize the distribution of variants or genomic features across the genome.
- **Expression Heatmaps:** In RNA-seq analysis, heatmaps are often used to visualize differential gene expression across multiple samples, clustering genes or samples based on their expression patterns.
- **Principal Component Analysis (PCA):** PCA can be used to visualize and assess the variance in the data, such as in differential gene expression or sample clustering.

Downstream Analysis and Interpretation

Once the raw data has been processed, variants have been called, and gene expression has been quantified, the final step is to interpret the results in the biological context. This may include:

- **Pathway Analysis**: Mapping differentially expressed genes or associated variants to known biological pathways using tools like **KEGG**, **Reactome**, or **Ingenuity Pathway Analysis** (IPA).
- **Functional Enrichment Analysis**: Identifying enriched biological functions, Gene Ontology (GO) terms, or protein-protein interaction networks in differentially expressed genes or mutated genes.
- **Clinical Interpretation**: In clinical genomics, the goal may be to identify pathogenic variants associated with diseases. Tools like **ClinVar**, **OMIM**, and **HGMD** help to classify variants based on clinical relevance.

NGS data analysis pipelines are essential for converting raw sequencing data into actionable biological insights. Each step in the pipeline—from quality control and alignment to variant calling and biological interpretation—requires careful attention and the use of specialized tools. The complexity and flexibility of NGS make these pipelines adaptable to a wide range of applications, including whole-genome sequencing, transcriptomics, and clinical diagnostics. With the continuous evolution of sequencing technologies and bioinformatics tools, NGS data analysis is becoming more efficient, accurate, and accessible, facilitating deeper insights into genomics, disease mechanisms, and personalized medicine.

APPLICATIONS OF NSG IN GENOMICS, METAGENOMICS AND EPIGENOMICS

NGS in Genomics

Genomics is the study of the complete genetic material of an organism, including its structure, function, evolution, and mapping. NGS has become the gold standard in genomics due to its ability to sequence entire genomes at a fraction of the cost and time of traditional methods, such as Sanger sequencing.

Applications in Genomics:

- **Whole-Genome Sequencing (WGS)**: NGS enables the sequencing of entire genomes, providing a comprehensive view of an organism's genetic makeup. Whole-genome sequencing (WGS) is used to identify genetic variants, such as single nucleotide polymorphisms (SNPs), insertions, deletions (indels), and structural variations. This approach is crucial for studying human genetic diseases, evolutionary biology, and population genetics. **Example**:

WGS is widely used in human genetics to identify rare genetic disorders, understand complex diseases like cancer, and investigate human evolution.

- **Exome Sequencing (WES):** While WGS sequences the entire genome, **Whole Exome Sequencing** (WES) focuses only on the coding regions of the genome (the exons), which constitute about 1-2% of the genome but harbor most disease-causing mutations. WES is more cost-effective than WGS while still providing valuable insights into genetic disorders. **Example:** WES is routinely used in clinical diagnostics to identify mutations in genes that cause inherited diseases.
- **Variant Calling and Annotation:** NGS platforms are used to detect and annotate genetic variants, which can range from common SNPs to rare mutations with clinical significance. Tools like GATK (Genome Analysis Toolkit) and ANNOVAR are widely used for variant calling and functional annotation, helping researchers link genetic variants to diseases. **Example:** Identification of somatic mutations in cancer genomes to guide targeted therapies.
- **De Novo Genome Assembly:** NGS technologies are essential for assembling genomes of organisms whose reference genomes are unavailable. **De novo sequencing** generates new genome assemblies from scratch by piecing together short sequencing reads. This application is especially useful for studying non-model organisms or pathogens. **Example:** De novo genome sequencing has been instrumental in sequencing the genomes of novel species, including agricultural crops, endangered species, and microbial pathogens.

NGS in Metagenomics

Metagenomics is the study of genetic material recovered directly from environmental samples, providing insights into the diversity, structure, and function of microbial communities. Unlike traditional microbiology techniques that focus on culturing individual species, metagenomics allows for the comprehensive study of all microorganisms within a given sample, including bacteria, viruses, fungi, and archaea.

Applications in Metagenomics:

- **Microbial Diversity Profiling:** NGS is used to study the diversity of microbial communities in various environments, including the human gut, soil, oceans, and extreme habitats. By sequencing the 16S rRNA gene (for bacteria and archaea) or the internal transcribed spacer (ITS) region (for fungi), NGS enables high-throughput identification and quantification of microbial taxa in complex samples. **Example:** NGS has been used to profile the human microbiome and its association with diseases such as obesity, inflammatory bowel disease, and autoimmune conditions.

- **Shotgun Metagenomics:** Shotgun metagenomic sequencing goes beyond the identification of species by sequencing the entire DNA content of a sample. This provides insights into the functional potential of microbial communities, including the genes responsible for metabolism, antibiotic resistance, and pathogenicity. Shotgun sequencing can also be used to identify novel microbes and their genetic properties. **Example:** Metagenomic shotgun sequencing has been used to study microbial communities in the gut, skin, respiratory tract, and in environmental samples, providing insights into their role in health and disease.
- **Antibiotic Resistance Gene Detection:** Metagenomic sequencing is a powerful tool for detecting antibiotic resistance genes (ARGs) in microbial communities, providing critical information about the spread of resistance in clinical, agricultural, and environmental settings. **Example:** NGS-based metagenomic studies have identified a wide range of novel antibiotic resistance genes in hospital wastewater, helping track the global rise of antimicrobial resistance.
- **Virus Discovery and Characterization:** NGS is an effective method for discovering novel viruses in environmental samples, particularly for those that are difficult to culture. By sequencing viral DNA or RNA directly from the sample, NGS can reveal new viral species, track viral evolution, and identify potential emerging pathogens. **Example:** NGS has been used to identify new viruses in bat populations, providing insights into zoonotic diseases like Ebola and SARS-CoV-2.

NGS in Epigenomics

Epigenomics is the study of the complete set of epigenetic modifications on the genetic material of a cell. These modifications regulate gene expression without altering the underlying DNA sequence and include DNA methylation, histone modifications, and non-coding RNA interactions. NGS has enabled a deeper understanding of the role of epigenetic changes in development, aging, disease, and environmental responses.

Applications in Epigenomics:

- **DNA Methylation Analysis:** DNA methylation is an important epigenetic modification that typically suppresses gene expression. NGS-based techniques, such as **Bisulfite Sequencing** (BS-seq) or **Methyl-Seq,** allow for high-resolution mapping of DNA methylation patterns across the entire genome. This is particularly useful for understanding gene regulation, imprinting disorders, cancer, and other diseases. **Example:** DNA methylation analysis using NGS has been used to identify biomarkers for early cancer detection, as tumors often exhibit distinct methylation patterns compared to normal tissue.

- **Histone Modification Mapping:** Histone modifications, such as acetylation and methylation, regulate chromatin structure and gene expression. NGS-based **ChIP-Seq (Chromatin Immunoprecipitation Sequencing)** is a powerful tool for identifying histone modifications and their association with transcriptional regulation, cell differentiation, and disease. **Example:** ChIP-Seq has been used to study how histone modifications are altered in diseases such as cancer, neurological disorders, and autoimmune diseases.
- **Non-Coding RNA Profiling:** NGS is widely used to analyze non-coding RNAs, such as microRNAs (miRNAs), long non-coding RNAs (lncRNAs), and small RNAs. These molecules play critical roles in the regulation of gene expression and cellular processes. Profiling non-coding RNAs using NGS helps elucidate their role in diseases like cancer, cardiovascular disease, and neurodegenerative disorders. **Example:** NGS-based profiling of miRNAs has been used to identify potential biomarkers for various cancers, including breast, lung, and prostate cancers.
- **Chromatin Accessibility (ATAC-Seq): ATAC-Seq (Assay for Transposase-Accessible Chromatin with sequencing)** is a technique that uses NGS to map regions of open chromatin, providing insights into transcription factor binding and the regulatory regions of the genome. This method is crucial for studying gene expression regulation in different cell types or conditions. **Example:** ATAC-Seq has been used to study the chromatin landscape of immune cells and to understand how chromatin remodeling influences gene expression in diseases such as leukemia.

13

Chapter

BIG DATA IN BIOINFORMATICS

INTRODUCTION

Big data in bioinformatics refers to the vast and complex datasets generated by modern biological research, which require advanced computational tools and techniques to analyze and interpret. With the advent of technologies like next-generation sequencing (NGS), high-throughput screening, and various omics approaches (genomics, proteomics, transcriptomics), biological data has grown exponentially in size and complexity. These massive datasets provide unprecedented insights into genetic variations, disease mechanisms, and biomolecular interactions, but they also pose significant challenges in terms of storage, processing, and analysis. Bioinformaticians use big data analytics to uncover patterns, make predictions, and derive meaningful biological knowledge, thereby driving advancements in personalized medicine, drug discovery, and disease prevention.

HIGH-THROUGHPUT DATA ANALYSIS

High-throughput data analysis refers to the computational methods and techniques used to handle and interpret large volumes of data generated by high-throughput technologies, such as next-generation sequencing (NGS), microarrays, and mass spectrometry. These technologies enable the simultaneous measurement of thousands or even millions of biological variables, such as genes, proteins, or metabolites, at once. While high-throughput approaches provide valuable insights into complex biological systems, they also present significant challenges in terms of data storage, quality control, and interpretation. To make sense of these massive datasets, bioinformaticians rely on advanced statistical methods, machine learning algorithms, and data visualization tools. These techniques allow for the identification of patterns, the discovery of new biomarkers, and the prediction of biological functions or disease outcomes. High-throughput data analysis is crucial in areas like genomics, transcriptomics, proteomics, and metabolomics, playing a key role in applications ranging from drug discovery to personalized medicine, where large-scale data integration and analysis are essential for making informed decisions. High-throughput data analysis is a critical component of modern bioinformatics and systems biology, driven by the need to analyze large, complex datasets generated by advanced laboratory technologies. These datasets typically come from high-throughput techniques like next-generation sequencing (NGS), microarrays, mass spectrometry, and other "omics" technologies, including genomics, transcriptomics, proteomics, and metabolomics. These technologies enable the collection of vast amounts of data, often in real-time, to measure thousands or even millions of biological variables (such as genes, proteins, metabolites, etc.) simultaneously.

Components of High-Throughput Data Analysis

Data Acquisition and Preprocessing

The first step in high-throughput data analysis is the acquisition of raw data, which is often noisy and incomplete. For example, in NGS, raw sequences are generated as long strings of nucleotides (A, T, C, G) that may contain errors. The data is then preprocessed to remove noise, correct errors, and standardize it for further analysis. Preprocessing may include:

- Quality control (QC): Removal of low-quality reads or sequences.
- Trimming and filtering: Removing adapter sequences or low-quality bases.
- Alignment: Aligning raw sequence data to a reference genome or transcriptome to determine the location of genes or mutations.

Data Storage and Management

Storing high-throughput data efficiently is a significant challenge due to its size. Typically, this data is stored in specialized formats (e.g., FASTQ for sequencing data, mzML for mass spectrometry data) and may require cloud-based or distributed storage solutions to handle the volume. Data management tools and databases (e.g., GenBank, ENCODE, or custom-built databases) help organize and make the data accessible for analysis.

Data Analysis Methods

After preprocessing, data undergoes various computational analyses. Some common techniques include:

- **Gene Expression Analysis:** In transcriptomics, high-throughput RNA sequencing (RNA-Seq) provides data on gene expression levels across different conditions. Analysis typically involves differential expression analysis (identifying genes that are up- or down-regulated in response to different conditions), clustering, and pathway enrichment analysis.
- **Variant Calling:** In genomics, high-throughput sequencing is used to identify genetic variants such as single nucleotide polymorphisms (SNPs),

insertions, deletions (indels), and structural variations. Variant calling algorithms (e.g., GATK, SAMtools) help identify these variations by comparing sequencing data to a reference genome.

- **Proteomics:** Mass spectrometry-based techniques in proteomics generate data on protein abundance and post-translational modifications. The data is analyzed to identify proteins, their relative quantities, and any modifications like phosphorylation or glycosylation.
- **Metabolomics:** Similar to proteomics, metabolomics techniques (e.g., NMR, GC-MS, LC-MS) generate data on the metabolites present in biological samples. Statistical tools help identify metabolite concentrations, metabolic pathways, and changes under different biological conditions.
- **Machine Learning:** To handle the complexity of high-dimensional data, machine learning methods such as clustering, classification, and regression are increasingly applied. Techniques like k-means clustering, hierarchical clustering, principal component analysis (PCA), and random forests are used to identify hidden patterns, classify samples, or predict biological outcomes (e.g., disease susceptibility, treatment response).

Statistical Analysis and Interpretation

Statistical methods play a pivotal role in analyzing high-throughput data, helping to interpret results accurately. Key techniques include:

- Differential analysis: Statistical tests like t-tests or ANOVA are used to identify significant differences between conditions (e.g., comparing gene expression levels between disease and control samples).
- Multiple Testing Correction: Given the large number of tests involved (e.g., testing thousands of genes or proteins), methods like the Bonferroni correction or the Benjamini-Hochberg procedure are used to adjust for false positives.
- Pathway and Network Analysis: After identifying differentially expressed genes or proteins, these are mapped to biological pathways or networks to understand their role in cellular processes or diseases. Tools like Gene Ontology (GO) enrichment analysis, KEGG pathways, or Ingenuity Pathway Analysis (IPA) are frequently used.

Data Integration and Visualization

Integrating multiple types of high-throughput data (e.g., combining genomics, transcriptomics, and proteomics data) is essential for a comprehensive understanding of biological systems. Various tools help combine and analyze these different data types, often leading to the discovery of novel biomarkers or therapeutic targets. Visualization tools, such as heatmaps, volcano plots, or network graphs, are essential for presenting large-scale data in an accessible format, enabling researchers to derive meaningful biological conclusions.

Challenges in High-Throughput Data Analysis

- **Data Complexity and Dimensionality:** High-throughput datasets are often high-dimensional, meaning they contain a large number of variables (e.g., thousands of genes or proteins), which makes analysis complex. Reducing dimensionality without losing key biological information is a significant challenge.
- **Data Quality:** Raw data generated by high-throughput technologies often contains errors, biases, and noise, which need to be addressed during preprocessing. For example, sequencing errors, cross-contamination in samples, or instrument calibration issues can affect the accuracy of the data.
- **Data Interpretation:** Even after sophisticated analysis, the biological interpretation of results can be difficult. High-throughput data typically provides large amounts of information, but translating this into actionable biological insights requires domain knowledge, robust computational tools, and rigorous validation.
- **Scalability:** As high-throughput technologies advance, the volume of data continues to grow, requiring increasingly sophisticated computational resources. Managing, processing, and analyzing datasets of this scale often require specialized software and high-performance computing infrastructure.
- **Reproducibility and Validation:** Reproducibility is a key concern in high-throughput analysis. In many cases, the results generated need to be validated through independent experiments or through statistical validation techniques. Ensuring that high-throughput findings can be reproduced in other laboratories or using different datasets is crucial for scientific progress.

Applications of High-Throughput Data Analysis

- **Genomic Medicine:** High-throughput genomic technologies, such as whole-genome sequencing (WGS) and exome sequencing (WES), are increasingly used for personalized medicine. Analyzing genetic variations in patient samples can help predict disease risk, tailor treatments, and even identify novel therapeutic targets.

- **Cancer Research:** High-throughput data analysis is extensively used in cancer research to identify genetic mutations, gene expression changes, and protein markers associated with various cancers. This helps in identifying potential biomarkers for early diagnosis, prognosis, and targeted therapies.
- **Drug Discovery:** High-throughput screening (HTS) allows researchers to test large libraries of chemical compounds for activity against biological targets. Data analysis helps identify promising drug candidates, understand their mechanisms of action, and optimize their properties.
- **Microbiome Studies:** High-throughput sequencing technologies, like 16S rRNA sequencing, are used to profile the microbiome (the community of microorganisms living in and on the human body). Data analysis can reveal associations between microbiome composition and diseases such as obesity, diabetes, and autoimmune conditions.
- High-throughput data analysis is at the heart of modern bioinformatics, enabling researchers to process and interpret large-scale datasets with the goal of understanding complex biological systems. While the challenges of managing and analyzing such vast amounts of data are considerable, the rewards include breakthroughs in areas like personalized medicine, drug discovery, and disease biology. As technology continues to advance, the potential applications and impact of high-throughput data analysis will only continue to grow, driving future discoveries in life sciences.

CLOUD COMPUTING AND BIOINFORMATICS

Cloud computing has become a transformative technology in many fields, and bioinformatics is no exception. As the complexity and volume of biological data increase, researchers and institutions are increasingly turning to cloud computing to manage, process, and analyze these large datasets. Cloud computing offers scalable, flexible, and cost-efficient solutions for handling the massive amounts of data generated in bioinformatics, particularly with the rise of high-throughput technologies such as next-generation sequencing (NGS), genomics, transcriptomics, and proteomics. Below is a detailed description of how cloud computing integrates with bioinformatics and the benefits it provides to researchers in this field. Cloud computing is the delivery of computing services over the internet, providing on-demand access to a wide range of computing resources, including servers, storage, databases, networking, software, and analytics tools. These resources are hosted on remote servers managed by cloud service providers such as Amazon Web Services (AWS), Google Cloud Platform (GCP), Microsoft Azure, and others. Cloud computing follows a pay-as-you-go model, where users pay only for the resources they use, making it highly scalable and cost-effective.

Benefits of Cloud Computing in Bioinformatics

- **Scalability**: One of the most significant advantages of cloud computing is its scalability. Bioinformatics projects often require massive computational power for data analysis, and cloud platforms provide virtually unlimited resources to scale up or down based on the project's demands. This is particularly useful in bioinformatics, where the volume of data can vary significantly, from small experimental datasets to massive datasets generated by high-throughput sequencing technologies.
- **Data Storage and Management**: Cloud computing offers vast storage capabilities, which are essential for bioinformatics workflows that generate petabytes of data. Biological data from high-throughput technologies can be large and complex, and cloud platforms allow researchers to store this data securely without worrying about local hardware limitations. Additionally, cloud storage solutions are designed to scale with the data, ensuring that researchers only pay for what they use. Many cloud providers also offer backup and redundancy options to ensure data safety and integrity.
- **Collaboration and Accessibility**: Bioinformatics research often involves teams of scientists and researchers across different locations and institutions. Cloud computing allows for easy collaboration by providing shared access to datasets and analysis pipelines, eliminating the need for physical data transfers or the maintenance of separate servers. Researchers can access their data and analysis tools from anywhere in the world, facilitating seamless collaboration among global teams.
- **Computational Power and Speed**: Bioinformatics workflows often require substantial computational power for tasks such as sequence alignment, genome assembly, data mining, or protein structure prediction. Cloud computing platforms provide access to high-performance computing (HPC) resources, which allow bioinformaticians to run complex algorithms and process large datasets in parallel. For instance, Amazon EC2 instances or Google Compute Engine provide powerful virtual machines that can be scaled to meet the needs of a particular task, significantly speeding up computational workflows.
- **Cost Efficiency**: Traditional computational infrastructure requires significant upfront investment in hardware and ongoing maintenance costs. With cloud computing, bioinformaticians can access the required resources without the capital expense, and the pay-as-you-go model ensures that they only pay for the resources they use. This makes cloud computing an especially attractive option for academic institutions, small research labs, or startups that may not have the budget to invest in dedicated infrastructure.

- **Security and Compliance**: Security is a major concern when working with sensitive biological data, especially when it involves personal health information, such as genomics data in clinical studies. Leading cloud service providers offer robust security measures, including data encryption (both in transit and at rest), access controls, and authentication mechanisms. Many cloud platforms are compliant with regulations such as the Health Insurance Portability and Accountability Act (HIPAA) and General Data Protection Regulation (GDPR), which are critical in healthcare and clinical bioinformatics applications.
- **Data Integration and Big Data Analytics**: Bioinformatics involves integrating various types of data from different sources, such as genomic, transcriptomic, proteomic, and metabolomic data. Cloud computing facilitates the integration of multiple datasets into unified analysis pipelines. Cloud platforms also provide powerful big data analytics tools, such as Hadoop and Spark, which can be used to process and analyze large datasets efficiently.
- **Data Sharing and Publishing**: The cloud also facilitates the sharing of data with the scientific community. Researchers can store datasets in public cloud repositories or share access to private cloud storage for collaborative research. Many bioinformatics workflows involve open-access data, and cloud-based platforms provide easy methods for publishing results and sharing datasets via online repositories (e.g., GenBank, ENCODE, or the cloud-specific Google Cloud Storage and AWS S3).

Common Use Cases of Cloud Computing in Bioinformatics

- **Genomics and Sequence Analysis**: Genomic data analysis, such as whole-genome sequencing (WGS) or RNA sequencing (RNA-Seq), requires significant computational resources for tasks like sequence alignment, variant calling, and annotation. Cloud platforms provide the computational power to perform these analyses on a large scale. Popular cloud-based tools and pipelines, such as the Broad Institute's FireCloud or Google Genomics, allow users to process genomic data without needing local infrastructure.
- **Proteomics**: Mass spectrometry-based proteomics workflows often generate massive datasets that need to be processed and analyzed using complex algorithms. Cloud-based tools allow researchers to store, manage, and analyze proteomic data efficiently. For example, cloud platforms can be used to conduct large-scale protein identification, quantification, and pathway analysis.

- **Machine Learning and AI in Bioinformatics**: Cloud computing enables the use of machine learning (ML) and artificial intelligence (AI) techniques for predictive modeling, classification, and pattern recognition in bioinformatics. For instance, ML algorithms are widely used in genomics for predicting gene function, detecting genetic variations, or identifying disease biomarkers. Cloud services like Amazon SageMaker or Google AI Platform provide infrastructure for building and deploying ML models at scale.
- **Medical and Clinical Bioinformatics**: In clinical bioinformatics, cloud platforms are used to integrate patient data, perform variant interpretation, and support personalized medicine. Cloud computing enables the analysis of large clinical datasets, such as electronic health records (EHRs) and genomic data, and facilitates clinical decision-making. For example, cloud-based systems can be used to identify genetic mutations related to diseases, suggest potential treatments, or predict disease risk based on genomic and clinical information.
- **Metagenomics**: Metagenomic sequencing involves the study of microbial communities in a sample (e.g., human microbiomes, soil, or environmental samples). Cloud computing is particularly useful for storing and analyzing large amounts of metagenomic sequencing data. Bioinformaticians use cloud-based tools to process shotgun sequencing data, assemble genomes, and analyze microbial diversity at scale.
- **Bioinformatics as a Service (BaaS)**: Several cloud providers offer Bioinformatics as a Service, where bioinformaticians can access pre-configured software tools and pipelines through the cloud. This enables researchers to perform complex bioinformatics tasks without the need for specialized local infrastructure or technical expertise. Cloud services like DNAnexus, Seven Bridges, and Galaxy on AWS provide bioinformatics workflows for researchers, offering data processing, analysis, and visualization tools on-demand.

Challenges and Considerations

While cloud computing offers many advantages, there are also some challenges associated with its use in bioinformatics:

- **Data Privacy and Security**: While cloud platforms generally offer strong security features, sensitive biological data, particularly personal health data, requires extra caution. Researchers must ensure that the data is encrypted and comply with legal and ethical standards, such as obtaining informed consent for the use of personal data in clinical genomics studies.

- **Data Transfer and Bandwidth**: Bioinformatics datasets, particularly from genomic sequencing, can be extremely large, making data transfer to and from the cloud potentially slow and costly, especially when dealing with high-volume datasets.
- **Learning Curve**: While cloud computing platforms are becoming more user-friendly, there may still be a learning curve for bioinformaticians who are not familiar with cloud-based technologies. Researchers need to learn how to set up cloud environments, manage resources, and optimize workflows.
- **Cost Management**: While cloud computing can be cost-effective, it is essential for bioinformaticians to monitor resource usage carefully. Uncontrolled usage, such as running unnecessary virtual machines or using excessive storage, can lead to unexpected costs.

DATA STORAGE AND RETRIEVAL FOR OMICS DATA

Omics data, including genomics, transcriptomics, proteomics, metabolomics, and other related fields, is central to modern biological and medical research. These data types can be vast, multidimensional, and highly complex, often comprising large volumes of raw and processed data that need to be securely stored, efficiently managed, and easily retrieved for further analysis. Managing omics data involves not only storing the data but also ensuring its accessibility, integrity, security, and scalability over time. Here, we will describe in detail the key concepts, technologies, and challenges involved in the storage and retrieval of omics data.

Characteristics of Omics Data

Omics data is typically characterized by:

- **High Volume:** High-throughput technologies such as next-generation sequencing (NGS) or mass spectrometry generate huge datasets. For example, sequencing the entire human genome generates billions of base pairs, while transcriptomic data from RNA-Seq experiments can involve millions of reads per sample.
- **Complexity:** Omics datasets are often multidimensional, containing diverse types of information, such as gene expression levels, protein abundance, or metabolite concentrations. These data types are interrelated, and integrating them often requires sophisticated data management approaches.
- **Variety:** Omics data can come from different types of biological samples (e.g., blood, tissue, cells) and various experimental conditions, which leads to a wide variety of formats, structures, and file types.

- **High Velocity:** Data generation in omics studies is often continuous and ongoing, especially in longitudinal studies, making real-time storage and retrieval a crucial component of data infrastructure.

Data Storage for Omics Data

Due to the volume, complexity, and sensitivity of omics data, storage solutions need to be efficient, scalable, and secure. There are several key aspects to consider when storing omics data.

Storage Types and Approaches

- **Local Storage:** Historically, omics data has been stored on local servers or hard drives. While this approach can work for small datasets, it quickly becomes unsustainable as data volume grows. Local storage can also be prone to data loss due to hardware failure and may limit collaboration across multiple research teams.
- **Cloud Storage:** Cloud storage is becoming the standard for handling large-scale omics data. Cloud services like Amazon Web Services (AWS), Google Cloud Platform (GCP), and Microsoft Azure provide highly scalable, on-demand storage solutions. These platforms offer both block storage (for raw data) and object storage (for unstructured data like genomic sequences).
- **Benefits of cloud storage:**
    - Scalability: The ability to easily expand storage capacity as data volume grows.
    - Cost Efficiency: Pay-as-you-go pricing models allow researchers to only pay for the storage they use.
    - Access and Collaboration: Cloud storage enables data sharing and remote access from anywhere, which is essential for collaborative omics research.
    - Security: Leading cloud providers ensure data is encrypted and comply with regulatory standards like HIPAA and GDPR.
- **High-Performance Storage Systems (HPC Storage):** Omics data often requires substantial computational resources for processing. Therefore, storage systems integrated with high-performance computing (HPC) infrastructures, such as parallel file systems (e.g., Lustre, GPFS), are used to allow high-speed data retrieval and processing.
- **Distributed Storage:** For large, geographically dispersed teams, distributed storage systems (e.g., Hadoop Distributed File System [HDFS]) are used. These systems break down data into smaller chunks that are distributed

across multiple storage locations. Distributed storage is highly resilient and allows parallel access to large datasets, improving both performance and data availability.

Data Formats and File Systems

Omics data is often stored in specialized file formats that are optimized for the type of analysis being performed. Common formats include:

- FASTQ: A format used to store raw sequence data from NGS, including base calls and associated quality scores.
- BAM/SAM: Formats for storing sequence alignment data, often used after NGS data has been aligned to a reference genome.
- VCF (Variant Call Format): Used to store genetic variant data (e.g., SNPs, indels) that have been identified in sequencing data.
- GFF/GTF: Formats used to store gene annotation data, describing the locations of genes, exons, and other elements in a genome.
- CSV/TSV: Tabular formats commonly used for storing gene expression or proteomics data.
- HDF5: A binary format for storing large datasets, particularly used in proteomics, metabolomics, and other high-throughput technologies.
- Specialized databases such as GenBank, Ensembl, and dbSNP are used to store genomic data, while proteomics data might be stored in repositories like PRIDE or PeptideAtlas.

Metadata Management

Omics data does not exist in isolation, and understanding the context of the data is crucial. Metadata refers to the information about the data, such as:

- Experimental conditions
- Sample preparation methods
- Biological context (species, organ, disease state)
- Data quality measures (e.g., read depth, alignment metrics)

Effective metadata management allows researchers to interpret omics data correctly and enhances the ability to find, share, and reuse data across different studies.

Data Retrieval for Omics Data

Efficient retrieval of omics data is as important as its storage. Retrieving large, complex datasets quickly and accurately requires specialized tools and methods.

Database Management Systems (DBMS)

- **Relational Databases (RDBMS):** Traditional databases like MySQL and PostgreSQL have been used for storing structured omics data, especially for smaller datasets. However, they can become inefficient when dealing with highly complex or large-scale omics data.
- **NoSQL Databases:** NoSQL databases such as MongoDB and Cassandra are increasingly used to store omics data, as they can handle large, unstructured, and semi-structured data with high scalability. These databases are often used for storing raw sequence data or large gene expression datasets that do not fit neatly into traditional relational models.
- **Specialized Bioinformatics Databases:** Many omics data types have specialized databases for storing and querying data. For example:
    - Genomics: Databases like Ensembl, UCSC Genome Browser, and GenBank store annotated genome sequences and associated data.
    - Proteomics: Databases like UniProt, PeptideAtlas, and PRIDE focus on storing protein sequences, modifications, and other proteomic information.
    - Metabolomics: Databases like HMDB (Human Metabolome Database) store detailed information about metabolites and metabolic pathways.
- **Data Warehouses:** Large-scale omics data repositories often use data warehouses to store integrated and preprocessed data, allowing for faster queries and access. For example, the Cancer Genome Atlas (TCGA) integrates genomic, transcriptomic, and clinical data from thousands of cancer samples.

Data Indexing and Search Engines

Efficient retrieval often depends on how well data is indexed. Omics databases and repositories use indexing techniques such as full-text indexing, hashing, and tree-based indexing (e.g., B-trees) to improve query performance.

- **ElasticSearch:** A distributed search engine often used in bioinformatics for indexing large volumes of biological data and enabling fast, complex searches.
- **Graph Databases:** Tools like Neo4j allow for querying interconnected omics data, such as genetic networks or metabolic pathways.

Data Querying and Analysis Tools

To facilitate the retrieval of specific subsets of data, bioinformaticians rely on a variety of query and analysis tools. These tools include:

- **SQL Queries:** Standard SQL queries are used for retrieving structured data from relational databases.
- **GraphQL:** A flexible query language for APIs, often used to retrieve complex biological datasets from web-based repositories or APIs (e.g., querying genomic variant data from public repositories).
- **Bioinformatics Software:** Tools like Galaxy, Ensembl VEP (Variant Effect Predictor), and IGV (Integrative Genomics Viewer) provide powerful interfaces for querying, visualizing, and analyzing omics data stored in databases or cloud-based storage systems.

Challenges in Data Storage and Retrieval for Omics Data

- **Data Volume:** The sheer size of omics datasets (e.g., whole-genome sequencing data can range from several gigabytes to terabytes per study) requires substantial infrastructure and advanced storage systems.
- **Data Integration:** Omics data comes from multiple sources and in multiple formats, making integration and standardization challenging. Efficient data retrieval depends on proper integration and harmonization across different datasets and databases.
- **Interoperability:** Omics data often needs to be accessed and analyzed across different platforms. Ensuring interoperability between databases and data formats is a key challenge in bioinformatics.
- **Data Security and Privacy:** Sensitive patient information in clinical omics studies needs to be securely stored and retrieved in compliance with privacy regulations (e.g., HIPAA, GDPR). Data encryption and proper access controls are necessary to ensure data protection.
- **Data Provenance:** It is crucial to track the history of omics data (from collection to analysis) to ensure its quality and reproducibility. Managing data provenance is a complex task that involves storing information about the source, transformations, and analysis steps applied to the data.

14 Chapter | PERSONALIZED MEDICINE AND BIOINFORMATICS

INTRODUCTION

Personalized medicine and bioinformatics represent two interconnected and transformative fields that are revolutionizing healthcare by enabling more tailored, precise, and effective treatments. Personalized medicine is an approach to medical treatment and healthcare that takes into account individual differences in patients' genes, environments, and lifestyles. By leveraging this information, healthcare providers can offer treatments that are specifically suited to each patient, improving outcomes and minimizing the risk of adverse effects. Bioinformatics plays a pivotal role in this evolution by utilizing computational techniques to analyze and interpret complex biological data, particularly genomic information. With the rapid advancement of technologies such as next-generation sequencing, bioinformatics allows for the identification of genetic variations, disease biomarkers, and potential therapeutic targets. Together, these fields hold the promise of not only improving the accuracy and efficiency of medical interventions but also transforming the way we understand and treat diseases, offering a more holistic, data-driven approach to medicine that has the potential to reshape healthcare on a global scale.

PRECISION MEDICINE AND GENOMIC MEDICINE

Precision Medicine and Genomic Medicine are two closely related concepts that represent the forefront of modern healthcare, aiming to tailor medical treatment and preventive strategies to individual patients based on their genetic makeup, environmental factors, lifestyle, and other personal characteristics. While these terms are often used interchangeably, they have distinct nuances.

Precision Medicine

Precision medicine refers to a broad, individualized approach to medical care that involves understanding and integrating a patient's unique characteristics, including their genetic profile, lifestyle choices, environment, and health history, to guide decision-making for disease prevention, diagnosis, and treatment. The goal is to move away from the traditional "one-size-fits-all" approach, where treatments are standardized for everyone, regardless of personal variability. Instead, precision medicine aims to customize healthcare to the individual to improve treatment effectiveness, minimize side effects, and provide the best possible outcomes.

Components of precision medicine include:

- **Genetic Information:** Precision medicine often uses genetic testing to understand how genes influence disease susceptibility and response to medications. By identifying genetic variations, clinicians can predict how individuals will respond to specific drugs, which allows them to choose the most effective therapies and avoid those that might cause adverse reactions.
- **Lifestyle and Environmental Factors:** In addition to genetics, precision medicine incorporates an understanding of how environmental exposures (e.g., pollutants, diet, stress) and lifestyle choices (e.g., exercise, smoking) affect health. For instance, some individuals may have genetic predispositions to certain diseases that can be exacerbated by their environment or lifestyle choices.
- **Personalized Therapies:** In oncology, for example, precision medicine has led to the development of targeted therapies that specifically address the molecular drivers of cancer in an individual's tumor, as opposed to conventional treatments like chemotherapy that affect both healthy and cancerous cells indiscriminately. This minimizes collateral damage and maximizes therapeutic efficacy.
- **Data-Driven Decision Making:** Precision medicine also relies heavily on large-scale data collection, including genomic data, medical records, and lifestyle data, which is analyzed using bioinformatics tools and artificial intelligence to guide treatment decisions.

Genomic Medicine

Genomic medicine is a subset of precision medicine that focuses specifically on the role of the genome—the complete set of DNA, including all of its genes—in health and disease. Genomic medicine uses detailed genomic data, typically derived from genomic sequencing (e.g., whole-genome sequencing, exome sequencing), to better understand the genetic basis of disease and its inheritance patterns. It allows healthcare providers to make more accurate diagnoses, identify individuals at risk for certain diseases, predict disease progression, and guide the selection of appropriate treatments.

Key aspects of genomic medicine include:

- **Genetic Diagnostics:** One of the core components of genomic medicine is the ability to diagnose genetic disorders through molecular testing.

Advances in next-generation sequencing (NGS) technologies allow for the identification of genetic mutations and variations that can lead to diseases, some of which may not be apparent through traditional diagnostic methods. For example, genomic testing is used to identify inherited conditions like cystic fibrosis or sickle cell anemia and can also help identify carriers of genetic diseases.

- **Pharmacogenomics:** This field, a key component of genomic medicine, studies how genetic variations affect an individual's response to medications. By understanding a patient's genetic makeup, doctors can predict how they will metabolize drugs, which drugs will be most effective, and which may cause harmful side effects. Pharmacogenomics is already used in clinical practice to guide dosing for drugs like warfarin (an anticoagulant) and certain cancer therapies.
- **Cancer Genomics:** In oncology, genomic medicine has had a particularly profound impact. Tumors are often caused by mutations in specific genes, and identifying these mutations through genomic profiling allows for personalized treatment options, such as targeted therapies that attack cancer cells while sparing healthy tissue. Genomic medicine also plays a role in early cancer detection, by identifying genetic risk factors that may predispose an individual to develop specific types of cancer, even before symptoms appear.
- **Gene Therapy:** Genomic medicine also encompasses the development of gene therapies, where specific genetic material is introduced into a patient's cells to treat or cure diseases. This approach holds promise for treating genetic disorders at their source, potentially reversing conditions like Duchenne muscular dystrophy or certain types of inherited blindness.
- **CRISPR and Genome Editing:** Genomic medicine is further enhanced by breakthroughs in genome editing technologies like CRISPR-Cas9, which allows for the precise alteration of genetic material. Researchers are exploring ways to correct genetic defects directly in patients' DNA, offering the possibility of curing hereditary diseases at the genetic level.

Differences Between Precision Medicine and Genomic Medicine

While **genomic medicine** focuses specifically on the genetic information of an individual, **precision medicine** takes a broader view by integrating genetic data with other personal factors, including lifestyle and environmental factors. Genomic medicine is an essential tool within precision medicine but is not synonymous with it.

- **Precision Medicine:** A comprehensive approach to healthcare that includes genetic, environmental, and lifestyle factors to personalize treatment and care.
- **Genomic Medicine:** A subset of precision medicine that specifically focuses on how genes and genetic variations impact health and disease.

Applications and Impact

Prevention: With knowledge of genetic predispositions, doctors can offer more targeted prevention strategies. For example, individuals with a family history of breast cancer who carry BRCA1 or BRCA2 mutations may consider preventive measures such as increased screening, prophylactic surgeries, or preventive therapies.

- **Treatment:** By incorporating genomic data, clinicians can select treatments that are tailored to an individual's genetic profile, leading to better outcomes. For example, in cancer treatment, genomic testing can identify mutations that drive the cancer, enabling the use of targeted therapies that specifically inhibit these mutations.
- **Drug Development:** Both genomic and precision medicine are accelerating drug discovery, helping pharmaceutical companies design drugs that target specific molecular pathways implicated in disease. This can lead to the development of more effective treatments with fewer side effects, as therapies are designed for specific patient subgroups rather than broad populations.
- **Ethical Considerations:** As with any cutting-edge medical approach, both precision and genomic medicine raise ethical concerns. These include issues related to privacy (e.g., who owns and has access to genetic data), the potential for genetic discrimination, and the psychological impact of genetic information (e.g., how individuals cope with knowing they have a high genetic risk for a certain disease).

Both precision medicine and genomic medicine are revolutionizing healthcare by providing personalized, data-driven approaches to treatment and prevention. As technologies continue to advance, the integration of genetic, environmental, and lifestyle data promises to further improve the precision and efficacy of medical interventions, ultimately leading to more effective treatments, better patient outcomes, and a deeper understanding of the complex mechanisms underlying health and disease. However, for these advances to reach their full potential, ongoing efforts are needed to address challenges such as data privacy, equitable access to healthcare, and the need for better healthcare infrastructure to support the integration of genomics into clinical practice.

BIOINFORMATICS IN DRUG DISCOVERY

Bioinformatics plays an essential role in modern drug discovery by enabling the use of computational tools and large-scale biological data to identify novel drug targets, optimize drug candidates, and predict the safety and efficacy of potential therapeutic agents. With the exponential growth in biological data—especially genomic, proteomic, and metabolomic data—bioinformatics has become a cornerstone in the pharmaceutical industry, accelerating the drug development process and enhancing the likelihood of success. In drug discovery, bioinformatics integrates data analysis, modeling, and simulation techniques to address key challenges such as identifying disease biomarkers, screening compounds, and optimizing clinical trial designs.

Roles of Bioinformatics in Drug Discovery

Target Identification and Validation

One of the first and most critical steps in drug discovery is identifying and validating the right biological target, typically a protein or nucleic acid, that plays a key role in the disease process. Traditional target identification often relied on experimental methods, but bioinformatics has revolutionized this step by allowing researchers to mine large datasets, including genomic sequences, transcriptomic data, and proteomic profiles, to identify potential targets more efficiently.

- **Genomic Data Mining:** Bioinformatics tools enable the analysis of large-scale genomic datasets to identify genetic mutations or variations associated with disease. For example, gene expression profiling using RNA-Seq can reveal overactive or mutated genes in cancer, making them potential drug targets.
- **Protein-Protein Interaction (PPI) Networks:** Bioinformatics approaches are used to map interactions between proteins, which can help identify key molecular pathways involved in disease. Disrupting a critical PPI can be a therapeutic strategy.
- **Pathway Analysis:** Bioinformatics tools can analyze metabolic and signaling pathways to identify druggable targets. Pathway enrichment analysis helps prioritize targets based on their involvement in disease mechanisms.

High-Throughput Screening (HTS) and Virtual Screening

Traditional high-throughput screening (HTS) involves testing thousands of chemical compounds in the lab to identify potential drug candidates. However, this process is time-consuming and expensive. Bioinformatics offers virtual screening methods to significantly reduce the number of compounds that need to be tested experimentally.

- **Molecular Docking:** Computational docking is a bioinformatics technique used to predict how small molecules (potential drugs) will interact with a target protein. The goal is to "dock" the drug to the target's active site and evaluate the binding affinity, allowing researchers to virtually screen large compound libraries and prioritize those most likely to succeed in vitro.
- **Ligand-Based Screening:** This approach uses known ligands (small molecules that bind to a target) to find similar compounds from a database. Bioinformatics tools help identify compounds with similar chemical structures or binding profiles.
- **Structure-Based Drug Design (SBDD):** This involves creating a 3D model of the target protein's structure and using it to design drugs that specifically interact with the protein. If the 3D structure of the target is known (via X-ray crystallography or NMR), bioinformatics methods can simulate how different compounds might interact with the target.

Drug Design and Optimization

Once a potential drug has been identified, bioinformatics tools are used to optimize its properties, such as binding affinity, selectivity, and drug-like characteristics (e.g., solubility, stability, and absorption). This phase involves:

- **Quantitative Structure-Activity Relationship (QSAR):** QSAR modeling is a computational method that correlates the chemical structure of a compound with its biological activity. By analyzing large datasets of compounds and their known activities, bioinformatics tools can predict the activity of new compounds, guiding the optimization of lead candidates.
- **Pharmacophore Modeling:** This technique involves identifying the key structural features of molecules that are necessary for interacting with a particular target. Bioinformatics can model these features to design new compounds that fit the pharmacophore and enhance the likelihood of a successful interaction with the target protein.
- **Lead Optimization:** After identifying a lead compound, bioinformatics tools can predict modifications to improve properties like potency, selectivity, and ADMET (Absorption, Distribution, Metabolism, Excretion, Toxicity) profiles. This stage is essential to reduce the likelihood of drug failures in later phases of development.

ADMET Prediction

ADMET properties—Absorption, Distribution, Metabolism, Excretion, and Toxicity—are critical to assessing the safety and efficacy of a drug. Bioinformatics has greatly advanced the ability to predict ADMET characteristics early in the drug discovery process, reducing the need for expensive and time-consuming in vivo studies.

- **Toxicity Prediction:** Using large datasets of known toxic compounds, bioinformatics tools can predict the toxicity of new drug candidates. Models are built to assess the likelihood of hepatotoxicity, cardiotoxicity, or other adverse effects based on chemical structure and molecular interactions.
- **Pharmacokinetics (PK) Modeling:** Computational methods can simulate how a drug is absorbed, distributed, metabolized, and excreted in the body. These predictions help guide the optimization of drug candidates for favorable PK profiles.

Biomarker Discovery

Biomarkers are measurable indicators of a biological state, such as proteins, genes, or metabolites associated with a disease. In drug discovery, identifying biomarkers can be pivotal in diagnosing diseases, monitoring drug responses, and predicting clinical outcomes.

- **Omics Data Analysis:** Bioinformatics tools can analyze genomics, transcriptomics, proteomics, and metabolomics data to discover biomarkers that are associated with diseases, drug responses, or treatment resistance. For example, cancer biomarker discovery is a major area where bioinformatics tools are used to identify specific mutations or protein expressions that can guide treatment decisions.
- **Biomarker Validation:** Once potential biomarkers are identified, bioinformatics tools can be used to validate their utility and reproducibility, ensuring they are reliable for use in clinical settings.

Systems Biology and Drug Repurposing

Systems biology involves the integration of complex biological data from multiple sources to understand how different components of a biological system (e.g., genes, proteins, and metabolites) interact. This holistic approach helps identify new therapeutic strategies and can aid in drug repurposing.

- **Drug Repurposing:** Bioinformatics tools can predict whether existing drugs, originally developed for one condition, might be effective for treating other diseases. By analyzing molecular signatures, disease pathways, and drug-target interactions, bioinformatics can uncover new uses for approved drugs, thus shortening development timelines and reducing costs.
- **Multi-Omics Integration:** Combining genomic, transcriptomic, proteomic, and metabolomic data allows bioinformatics tools to generate a systems-level understanding of disease biology. This can identify novel therapeutic

targets and enable the development of more personalized treatment strategies.

Clinical Trial Design and Monitoring

Bioinformatics also plays an important role in the clinical development phase, particularly in designing more efficient clinical trials and monitoring patient responses.

- **Patient Stratification:** By analyzing genomic and clinical data, bioinformatics can help stratify patients based on their genetic makeup, disease subtype, or predicted response to a treatment. This allows for more personalized clinical trials, improving the chances of success and minimizing trial costs.
- **Clinical Data Integration:** Bioinformatics tools are used to integrate and analyze data from clinical trials, enabling better monitoring of patient outcomes, side effects, and biomarkers of response. This data can guide further optimization of drug candidates or adjust treatment protocols in real time.

Bioinformatics has become an indispensable part of the drug discovery process, dramatically increasing the speed, accuracy, and success rate of new drug development. By enabling more efficient target identification, drug screening, optimization, and safety profiling, bioinformatics tools are helping researchers navigate the complexities of biology and the vast amounts of data that are now available. As technology continues to advance, the role of bioinformatics in drug discovery will only grow, driving innovations in drug development and leading to more effective, personalized treatments for a wide range of diseases. The integration of computational models, high-throughput screening, multi-omics data, and systems biology approaches has the potential to transform how drugs are discovered, developed, and brought to market, ultimately improving patient care and therapeutic outcomes.

PHARMACOGENETICS AND DISEASE PREDICTION

Pharmacogenetics is a branch of pharmacology that studies the role of genetic variation in drug response. It focuses on how an individual's genetic makeup influences their response to medications, including efficacy, adverse effects, and optimal dosing. By integrating pharmacogenetic data into clinical practice, healthcare providers can tailor drug treatments to the genetic profiles of patients, leading to more effective therapies and minimizing harmful side effects. On the other hand, disease prediction refers to the identification of individuals at high risk for certain diseases based on their genetic, environmental, and lifestyle factors. The application of pharmacogenetics in disease prediction is transforming personalized medicine, providing more precise tools for both prevention and treatment.

Pharmacogenetics in Drug Response

Pharmacogenetics examines the genetic factors that contribute to interindividual variability in drug response, which can be influenced by:

Genetic Variants in Drug Metabolism

One of the key ways in which pharmacogenetics influences drug response is through variations in genes that encode drug-metabolizing enzymes. The most well-studied genes in this context are the cytochrome P450 (CYP) family of enzymes, which are responsible for metabolizing a wide variety of drugs. Variants in these genes can lead to differences in drug metabolism speed, which impacts drug levels in the body.

- **Ultra-rapid Metabolizers:** Some individuals have genetic variants that cause them to metabolize drugs more quickly than average, potentially leading to reduced drug effectiveness because the drug is eliminated too fast.
- **Poor Metabolizers:** Others may have genetic variations that cause slower drug metabolism, increasing the risk of drug toxicity because the drug stays in the body for too long at high concentrations.
- **CYP2D6:** One of the most extensively studied genes in pharmacogenetics is CYP2D6, involved in the metabolism of many drugs, including antidepressants, beta-blockers, and opioids. Individuals with mutations in this gene may have altered responses to these medications.

Drug Transporters and Receptors

In addition to metabolic enzymes, drug transporters and drug receptors play an important role in determining how a drug is absorbed, distributed, and acted upon in the body. Variations in these genes can lead to differences in drug response.

- **SLCO1B1:** A common genetic variant in the SLCO1B1 gene, which encodes a liver transporter, affects the way statins (cholesterol-lowering drugs) are processed. Some variants increase the risk of statin-induced muscle toxicity, prompting the need for dose adjustments or alternative medications.
- **Beta-adrenergic Receptors:** Variations in genes encoding beta-receptors can impact the efficacy of beta-blockers, commonly prescribed for conditions like hypertension or heart failure. A patient's genetic profile can guide doctors in selecting the most effective drug or adjusting the dose to ensure optimal results.

Pharmacodynamics and Drug Efficacy

Pharmacogenetics also includes the study of genetic variants that affect how drugs interact with their molecular targets—such as receptors, enzymes, or ion channels. These variations can influence drug efficacy.

- **Tamoxifen and Estrogen Receptor:** Tamoxifen is used to treat hormone-receptor-positive breast cancer. It is metabolized into its active form by the CYP2D6 enzyme. However, individuals with certain genetic variants in CYP2D6 may not metabolize tamoxifen effectively, leading to reduced treatment efficacy and a higher risk of cancer recurrence.

Adverse Drug Reactions (ADRs)

Genetic factors can also predict an individual's susceptibility to adverse drug reactions, some of which can be life-threatening. For example, the HLA-B*1502 allele, found more commonly in individuals of Asian descent, is associated with an increased risk of severe skin reactions to carbamazepine, an anticonvulsant drug. By testing for this genetic variant before prescribing carbamazepine, doctors can avoid these severe reactions.

- **Warfarin Sensitivity:** Warfarin, an anticoagulant, requires careful dosing due to its narrow therapeutic range. Variants in the VKORC1 and CYP2C9 genes can influence warfarin sensitivity. Patients with certain variants may require lower doses to avoid bleeding complications, and pharmacogenetic testing can guide proper dosing.

Disease Prediction and Genetic Risk Factors

Disease prediction using genetic information involves identifying individuals at high risk for certain diseases based on inherited genetic variations. Genetic testing can detect mutations or genetic predispositions that increase the likelihood of developing diseases such as cancer, cardiovascular disease, diabetes, or neurological disorders.

Cancer Risk Prediction

Genetic testing for mutations in oncogenes or tumor suppressor genes can help identify individuals at higher risk of developing certain types of cancer.

- BRCA1/BRCA2 Mutations: Mutations in the BRCA1 and BRCA2 genes are strongly associated with an increased risk of breast and ovarian cancers. Genetic screening for these mutations can help identify individuals who may benefit from enhanced surveillance, preventative measures (like mastectomy or oophorectomy), or targeted therapies such as PARP inhibitors.

- Lynch Syndrome: Individuals with Lynch syndrome have a higher risk of developing colorectal, endometrial, and other cancers due to inherited mutations in mismatch repair genes. Early genetic testing can lead to more frequent screenings and preventive strategies to reduce cancer risk.

Cardiovascular Disease Prediction

Certain genetic variants have been identified that increase susceptibility to cardiovascular diseases, such as coronary artery disease, hypertension, and hyperlipidemia.

- Familial Hypercholesterolemia (FH): FH is caused by mutations in the LDLR gene, leading to high cholesterol levels from a young age and a significantly increased risk of early heart disease. Genetic screening for FH can help identify at-risk individuals who may benefit from early intervention, lifestyle changes, and cholesterol-lowering treatments.
- Apolipoprotein E (APOE): Variants of the APOE gene are associated with an increased risk of Alzheimer's disease, as well as cardiovascular diseases. APOE testing can provide information about an individual's genetic risk, although environmental factors still play a major role in disease development.

Neurodegenerative Diseases

Several genetic mutations are linked to neurodegenerative diseases such as Parkinson's disease, Alzheimer's disease, and Huntington's disease.

- **Huntington's Disease:** Huntington's disease is caused by an expansion of CAG repeats in the HTT gene. Genetic testing can confirm a diagnosis and predict when symptoms might begin, although there is currently no cure for the disease.
- **Alzheimer's Disease:** The APOE ε4 allele is associated with an increased risk of Alzheimer's disease. Genetic testing for this allele can help predict risk, although it is not deterministic, and many individuals with the allele may never develop Alzheimer's.

Diabetes Prediction

Diabetes, especially Type 2 diabetes, has a strong genetic component. The identification of genetic variants that predispose individuals to diabetes can lead to early intervention strategies.

- **Type 1 Diabetes:** Genetic variations in the HLA gene complex can increase the risk of developing Type 1 diabetes. Although environmental factors like viral infections are also implicated, genetic testing can help identify those at higher risk.

- **Type 2 Diabetes:** Numerous genetic variants, such as those in the TCF7L2 gene, have been associated with an increased risk of Type 2 diabetes. Genetic screening, combined with lifestyle interventions, can help in early detection and prevention strategies.

Pharmacogenetics and Disease Prediction in Personalized Medicine

The combination of pharmacogenetics and disease prediction creates a powerful tool for personalized medicine—an approach that tailors medical treatment to the individual's genetic profile. By knowing a patient's genetic risk for certain diseases and how they respond to various drugs, healthcare providers can optimize prevention, diagnosis, and treatment.

- **Personalized Treatment Plans:** By integrating pharmacogenetic testing into routine clinical practice, doctors can prescribe drugs that are most likely to be effective for the individual, reducing the trial-and-error approach traditionally used in medicine.
- **Risk Stratification:** Genetic testing can stratify patients into different risk categories for various diseases. For example, a patient with a family history of heart disease and certain genetic markers can be closely monitored for early signs of cardiovascular disease and receive appropriate interventions.
- **Preventative Strategies:** For individuals identified as having a genetic predisposition to certain diseases, lifestyle changes (diet, exercise, smoking cessation) and preventive treatments (medications, vaccines) can be implemented earlier, potentially preventing or delaying disease onset.

Pharmacogenetics and disease prediction are central to the future of precision medicine, providing individualized, data-driven insights into drug therapy and disease risk. By understanding the genetic basis of drug responses and disease susceptibility, healthcare providers can make more informed decisions, improve treatment outcomes, and reduce the risk of adverse events. With the ongoing advancement of genomic technologies, the ability to predict, prevent, and treat diseases based on genetic information will continue to evolve, ultimately leading to a more personalized and effective approach to healthcare.

15 Chapter

BIOINFORMATICS FOR CANCER GENOMICS

INTRODUCTION

Bioinformatics for cancer genomics is a rapidly evolving field that leverages computational tools and techniques to analyze vast amounts of genetic data, providing critical insights into the molecular basis of cancer. By integrating high-throughput sequencing technologies, such as next-generation sequencing (NGS), with advanced data analysis platforms, bioinformaticians can identify mutations, gene expression patterns, and genomic alterations that drive tumor development and progression. This multidisciplinary approach not only enhances our understanding of cancer biology but also holds the potential to revolutionize cancer diagnostics, treatment, and personalized medicine. As the complexity of cancer genomes becomes increasingly evident, bioinformatics plays a crucial role in uncovering the genetic heterogeneity of tumors, guiding therapeutic decision-making, and enabling the development of targeted therapies and precision medicine strategies.

CANCER GENOME ATLAS

The Cancer Genome Atlas (TCGA) is a landmark project that has transformed our understanding of cancer genomics by providing comprehensive molecular profiles of various types of cancers. Launched in 2006 by the National Cancer Institute (NCI) and the National Human Genome Research Institute (NHGRI), TCGA aimed to explore the genetic underpinnings of cancer through high-throughput genomic sequencing and advanced bioinformatics. The project generated detailed genomic data, including DNA sequence mutations, gene expression patterns, copy number alterations, epigenetic modifications, and protein expression across a wide range of cancer types.

Characteristics Features of Cancer Genome Atlas (TCGA)

- **Comprehensive Cancer Genome Data:** TCGA collected data from over 11,000 cancer samples across more than 30 different cancer types, making it one of the most expansive cancer genomics datasets available. These data include both primary tumors and paired normal tissues, enabling comparisons between cancerous and non-cancerous genomes to identify specific genetic alterations that drive tumorigenesis.

- **Multi-dimensional Data:** The project focused on multi-dimensional genomic data, which includes:
    - **Somatic mutations**: Changes in DNA that occur in the tumor cells but not in the normal cells, such as single nucleotide variations (SNVs) and small insertions or deletions (indels).
    - **Copy number alterations (CNAs)**: Large-scale changes in the DNA, such as amplifications or deletions of entire chromosomal regions.
    - **Gene expression**: Patterns of RNA expression that indicate which genes are active or inactive in tumor cells, often linked to cancer progression.
    - **DNA methylation**: Epigenetic changes that can silence or activate genes without altering the underlying DNA sequence.
    - **Protein expression and post-translational modifications**: Proteomics data providing insight into how genetic changes are manifested at the protein level.
- **Data Access and Open Resources:** TCGA made its data publicly available through various platforms, including the Genomic Data Commons (GDC), enabling researchers worldwide to explore and analyze cancer-related genetic alterations. By making this data freely accessible, TCGA accelerated global cancer research, providing the foundation for a wide range of studies exploring cancer biology, diagnostics, and therapy development.
- **Identification of Cancer Drivers:** One of the most significant outcomes of TCGA was the identification of novel cancer driver genes and pathways. These are the genetic alterations that directly contribute to the initiation and progression of cancer. The project revealed recurrent mutations in well-known oncogenes (such as **TP53**, **KRAS**, **PIK3CA**, and **EGFR**) as well as previously unrecognized alterations in new cancer-related genes, thereby providing valuable targets for further investigation and therapeutic development.
- **Subtypes and Tumor Heterogeneity:** TCGA also revealed the remarkable heterogeneity within and between tumor types. It has provided evidence that even cancers of the same histological type can exhibit distinct genetic alterations, leading to the classification of cancers into molecular subtypes with different prognoses and treatment responses. For example, in breast cancer, TCGA helped define different subtypes such as basal-like, HER2-enriched, and luminal A and B, which differ not only in genetic alterations but also in clinical behavior.

- **Implications for Precision Medicine:** The data generated by TCGA has been instrumental in the development of precision medicine approaches in oncology. By identifying specific mutations and alterations in cancer genomes, TCGA data enables the development of targeted therapies aimed at those particular genetic changes. This approach holds the potential for more effective and less toxic treatments compared to traditional one-size-fits-all chemotherapy.
- **Collaborative Nature and Global Impact:** The TCGA project brought together researchers from around the world, fostering collaboration between academic, clinical, and governmental organizations. Its findings have spurred the development of numerous follow-up studies, clinical trials, and diagnostic tests aimed at translating genomic discoveries into clinical practice.

Achievements of TCGA

- **Identification of actionable mutations**: TCGA's discoveries have led to the identification of genetic mutations that can be targeted with specific drugs, such as the mutation in **BRAF** in melanoma, which can be treated with BRAF inhibitors.
- **Molecular subtyping of cancers**: TCGA helped define molecular subtypes of cancers, enabling more accurate prognostic predictions and personalized treatment plans.
- **Biomarker discovery**: TCGA has contributed to the discovery of potential biomarkers for early cancer detection, prognosis, and monitoring of treatment response.

Legacy and Ongoing Impact

Although the TCGA project officially concluded its major data collection in 2017, its legacy continues to shape cancer research. Many of the genomic data and findings from TCGA have been integrated into subsequent studies and clinical trials, furthering the understanding of cancer genomics and enhancing efforts to develop personalized cancer therapies. The success of TCGA has inspired similar initiatives worldwide, including the International Cancer Genome Consortium (ICGC), which continues to expand the catalog of cancer-associated genomic alterations. The Cancer Genome Atlas has been a transformative initiative in the field of cancer research, providing an unprecedented level of genomic insight into cancer biology. By enabling researchers to better understand the molecular foundations of cancer, TCGA has laid the groundwork for the development of more effective, personalized cancer treatments.

TUMOR MUTATION BURDEN AND NEOANTIGENS

Tumor Mutation Burden (TMB) and neoantigens are two closely related concepts in cancer immunology that have gained significant attention in recent years, particularly in the context of immunotherapy. TMB refers to the total number of somatic mutations within a tumor genome, while neoantigens are novel protein fragments generated by tumor-specific mutations, which can be recognized by the immune system as foreign or abnormal. Both TMB and neoantigens are critical factors influencing the effectiveness of immunotherapies, such as checkpoint inhibitors, and can provide important insights into tumor behavior, prognosis, and treatment response.

Tumor Mutation Burden (TMB)

Tumor Mutation Burden (TMB) refers to the total number of somatic mutations per megabase (mut/Mb) of the coding region of the tumor genome. These mutations arise during the process of tumorigenesis and accumulate over time. High TMB indicates a tumor has a large number of mutations, whereas low TMB suggests fewer genetic alterations. TMB can be measured through genomic sequencing techniques, such as whole-exome sequencing (WES) or targeted sequencing panels, which help identify point mutations, insertions, deletions, and other genetic alterations across the tumor genome.

Significance in Cancer

- **Correlating TMB with Immunotherapy Response:** High TMB is associated with a higher likelihood of generating neoantigens, which are potentially recognizable by the immune system. Tumors with a high TMB are often more immunogenic, meaning they are more likely to be targeted by immune cells such as T lymphocytes. This makes high TMB an important predictive biomarker for the response to **immune checkpoint inhibitors** (ICIs), such as **PD-1** and **CTLA-4** inhibitors.

  Several clinical studies have demonstrated that cancers with a high TMB, such as melanoma, non-small cell lung cancer (NSCLC), and bladder cancer, tend to respond better to checkpoint blockade therapies because the immune system is more likely to recognize and attack tumor cells.

- **TMB and Tumor Type:** Tumors with inherently high mutation rates tend to have higher TMBs. For example, cancers with a known mutagenic exposure (like lung cancer due to smoking, or melanoma due to UV radiation) often present with higher TMBs. On the other hand, cancers like breast cancer or ovarian cancer typically have lower TMB.
- **Thresholds for TMB:** The optimal TMB threshold for predicting response

to immunotherapy is still being studied, but research suggests that tumors with a TMB above a certain cutoff (e.g., >10-20 mutations/Mb) are more likely to respond to immune checkpoint inhibitors. Regulatory bodies like the **FDA** have approved TMB as a biomarker for some cancers, like NSCLC, where high TMB can justify the use of pembrolizumab (Keytruda) as first-line therapy.

- **Prognostic Value**: High TMB has been associated with worse prognosis in certain cancers, suggesting that higher mutation load may lead to increased tumor heterogeneity, potentially making treatment more difficult. However, high TMB also improves the prospects for immune system recognition, which can be leveraged therapeutically.

Neoantigens

Neoantigens are new antigens that arise from tumor-specific mutations. These antigens are formed when a tumor mutation leads to the production of abnormal or altered proteins, which are then processed and presented on the surface of tumor cells in the context of **major histocompatibility complex (MHC)** molecules. Neoantigens are recognized as «non-self» by the immune system, triggering an immune response.Neoantigens are unique to each individual›s tumor, as they arise from somatic mutations rather than inherited genetic variants. This makes neoantigens highly specific to the tumor and patient, representing a potential target for personalized cancer immunotherapies.

Mechanism of Neoantigen Generation

- **Mutation-Driven:** Neoantigens result from tumor-specific mutations in the DNA. These mutations may be point mutations, small insertions/deletions, or larger structural alterations, and can lead to altered protein sequences that are presented by MHC molecules on the tumor cell surface.
- **Processing and Presentation**: After a tumor cell produces a mutated protein, the protein is broken down into small peptides by the proteasome. These peptides are then loaded onto MHC class I (or class II) molecules and presented on the surface of the tumor cell, where they can be recognized by T-cells.
- **Immune Activation**: The presence of these foreign peptides on the tumor cell surface stimulates T-cells, particularly cytotoxic CD8+ T-cells, to recognize and attack the tumor cell.

Significance in Cancer Immunotherapy

- **Personalized Immunotherapy**: Since neoantigens are specific to individual tumors, they hold great promise for personalized cancer vaccines and immunotherapies. By identifying the specific neoantigens present in a patient's tumor, researchers can design personalized treatments that selectively target these neoantigens, minimizing damage to healthy tissue. Cancer vaccines, engineered T-cell therapies, and adoptive T-cell transfer are examples of approaches aiming to harness the power of neoantigen-targeting.

- **Immune Recognition and Tumor Evasion:** Not all mutations will result in neoantigens that can be effectively recognized by the immune system. Tumors can sometimes evolve mechanisms to evade immune detection by downregulating MHC expression or by inducing immune tolerance. Thus, the ability of a tumor to generate immunogenic neoantigens is influenced by both the nature of the mutations and the tumor's immune environment.
- **Neoantigen-Based Vaccines**: Tumor-specific neoantigen vaccines are an exciting area of research in cancer immunotherapy. These vaccines aim to stimulate a patient's immune system to recognize and attack tumor cells based on their unique neoantigen profile. Clinical trials are exploring neoantigen-based vaccines in various cancers, including melanoma, lung cancer, and glioblastoma.
- **Relationship Between TMB and Neoantigen Load**: There is a direct relationship between TMB and neoantigen load. Tumors with high TMB are more likely to generate a greater number of neoantigens, increasing the likelihood of the tumor being recognized and attacked by the immune system. As such, TMB serves as a surrogate marker for the potential abundance of neoantigens within a tumor.

Clinical Implications and Challenges

- **Biomarker for Immunotherapy Response**: TMB and neoantigen load are used as biomarkers for predicting which patients are most likely to benefit from immunotherapy, especially immune checkpoint inhibitors. However, the threshold for high TMB and neoantigen burden is still under investigation, and clinical application requires careful consideration of tumor type, mutation spectrum, and immune microenvironment.
- **Therapeutic Development**: Ongoing efforts are focused on developing therapies that can enhance the presentation and recognition of neoantigens. This includes cancer vaccines, engineered T-cell therapies (such as CAR-T cells), and neoantigen-based adoptive cell transfer.
- **Tumor Heterogeneity**: Tumor heterogeneity—where different parts of the same tumor or metastatic sites harbor distinct mutations—can complicate the neoantigen landscape. A therapeutic strategy that targets a specific set

of neoantigens might not be effective if the tumor harbors other mutations in different areas, highlighting the challenge of developing broadly effective treatments.

Tumor Mutation Burden (TMB) and neoantigens are integral components of the immune landscape in cancer. High TMB often correlates with an increased number of neoantigens, which can trigger immune responses against tumors. These biomarkers have opened new avenues for immunotherapy, especially in the context of immune checkpoint inhibitors and personalized immunotherapies targeting neoantigens. However, challenges such as tumor heterogeneity and immune evasion mechanisms need to be addressed to optimize the clinical utility of these markers in cancer treatment. As research progresses, TMB and neoantigen-based strategies are likely to become central pillars in precision oncology.

BIOINFORMATICS IN CANCER DRUG DEVELOPMENT

Bioinformatics has become an indispensable tool in cancer drug development, providing critical insights into the complex molecular mechanisms driving cancer and enabling more efficient, targeted therapeutic approaches. By leveraging computational tools, databases, and advanced analytics, bioinformaticians can analyze vast amounts of genomic, transcriptomic, proteomic, and clinical data to identify new drug targets, predict drug responses, and streamline the drug discovery process. This has led to the development of more precise, personalized therapies aimed at treating cancer more effectively while minimizing side effects.

BIOINFORMATICSISTRANSFORMINGCANCERDRUGDEVELOPMENT

Target Identification and Validation

The first step in cancer drug development is identifying molecules that play a central role in cancer progression, such as oncogenes, tumor suppressor genes, and other proteins that regulate cancer-related pathways. Bioinformatics approaches are widely used to identify new drug targets by analyzing large-scale genomic and transcriptomic datasets.

Genomic Data Analysis:

- High-throughput sequencing technologies, such as next-generation sequencing (NGS), provide large volumes of genomic data. Bioinformatics tools can analyze this data to identify somatic mutations (point mutations, indels, copy number alterations) and structural variations in cancer genomes that might affect critical genes and pathways.
- **Cancer Genomic Databases** like **The Cancer Genome Atlas (TCGA)** and **COSMIC** provide detailed mutation profiles across various cancer types, which help identify common and driver mutations.

Gene Expression Profiling:

- By analyzing RNA-Seq data, bioinformatics tools can identify overexpressed or underexpressed genes in tumors compared to normal tissue, highlighting potential targets for drug development.
- **Pathway Analysis:** Bioinformatics tools (e.g., **Gene Set Enrichment Analysis** or **Ingenuity Pathway Analysis**) can also be used to identify key pathways involved in tumorigenesis (such as **PI3K/AKT**, **MAPK**, or **p53 signaling pathways**) that could be targeted by drugs.

Target Validation:

- Once potential targets are identified, bioinformatics methods can be used to validate these targets by assessing their relevance to cancer biology and their potential druggability (i.e., how easy it is to design molecules that interact with the target).
- Systems biology approaches, including **network analysis**, help to understand the complex interactions between genes, proteins, and cellular pathways that control cancer cell behavior, validating whether a target is involved in tumor growth, invasion, and metastasis.

Drug Repurposing and Screening

Once a target is identified, the next step in cancer drug development is identifying compounds that can modulate the activity of the target. Bioinformatics plays a significant role in drug repurposing and virtual screening.

Drug Repurposing:

- Bioinformatics allows for the repurposing of existing drugs by analyzing their molecular profiles and exploring whether they may act on newly identified cancer targets. This approach reduces the time and cost involved in drug development. Databases like **DrugBank** and **Repurposing Drugs for Effective Anticancer Therapies (REACT)** provide comprehensive information on known drugs, including their mechanisms of action and molecular targets.

Virtual Screening and Molecular Docking:

- Bioinformatics tools such as **AutoDock**, **DockingServer**, or **Schrodinger's Glide** are used for virtual screening of compound libraries against target proteins. These tools simulate how small molecules interact with proteins at the molecular level, predicting which compounds may bind effectively to a target and exhibit biological activity.
- **Chemoinformatics** platforms, such as **PubChem** and **ChEMBL**, allow

researchers to analyze large chemical databases for known compounds that could serve as starting points for drug discovery.

Pharmacophore Modeling:

- **Pharmacophore modeling** is used to identify key structural features required for a compound to interact with a target protein. This approach aids in identifying new drug candidates that possess the essential structural properties for binding to the target.

Biomarker Discovery for Personalized Medicine

One of the greatest advancements in cancer drug development has been the move toward **personalized medicine**, where therapies are tailored to the specific genetic makeup of individual patients. Bioinformatics plays a critical role in discovering biomarkers that can predict how a patient will respond to a particular treatment.

Genomic Profiling for Patient Stratification:

- **Next-generation sequencing (NGS)** of tumors allows the identification of mutations, copy number variations, and gene expression profiles that could serve as predictive biomarkers for cancer treatments. For example, tumors with mutations in **EGFR**, **BRAF**, or **HER2** can be treated with targeted therapies like **EGFR inhibitors** (e.g., erlotinib), **BRAF inhibitors** (e.g., vemurafenib), or **HER2-targeted therapies** (e.g., trastuzumab).

Transcriptomic and Proteomic Profiling:

- **RNA-Seq** data can be analyzed to detect tumor-specific RNA signatures that correlate with specific cancer subtypes or drug responses, identifying patients who may benefit from a particular drug or treatment regimen.
- Similarly, proteomics (the study of proteins) helps in identifying biomarkers that predict tumor behavior or treatment resistance. Tools like **Mass Spectrometry** and bioinformatics platforms for **quantitative proteomics** (e.g., **MaxQuant**, **Proteome Discoverer**) allow the identification of differential protein expression in cancer cells and tissues.

Predictive Models for Drug Sensitivity:

- Machine learning and AI-driven bioinformatics models can predict how individual patients will respond to certain drugs based on their genetic profile. For example, tools like **Cancer Cell Line Encyclopedia (CCLE)** or **Genomics of Drug Sensitivity in Cancer (GDSC)** provide large datasets of drug responses across a variety of cancer cell lines, which can be used to predict therapeutic efficacy in patients with similar genetic alterations.

- **Clinical Trial Design and Patient Selection:** Bioinformatics tools also play a crucial role in designing clinical trials and selecting patients who are most likely to benefit from new cancer therapies.

Patient Cohort Stratification:

- Bioinformatics analyses of patient genomic data can be used to stratify patients into subgroups based on genetic alterations, ensuring that clinical trials include individuals who are most likely to respond to a particular therapy. This approach is critical for precision medicine and minimizes the risk of ineffective treatments.

Survival and Prognostic Modeling:

- Bioinformatics techniques such as **survival analysis** and **machine learning models** (e.g., **Cox proportional hazards model**, **random forests**) can help predict clinical outcomes, such as overall survival or progression-free survival, based on genetic and molecular data. This can guide the design of trials and help clinicians identify which patients are likely to benefit most from a new drug.

Real-Time Data Integration:

- Clinical trial data, including genomic and clinical variables, can be integrated and analyzed in real-time using bioinformatics tools. This allows for adaptive trial designs, where trial protocols can be adjusted based on interim results or emerging patient responses.

Mechanisms of Drug Resistance

Understanding mechanisms of drug resistance is essential for developing next-generation cancer therapies. Tumors can evolve resistance to therapies through genetic mutations, epigenetic changes, or the activation of compensatory signaling pathways. Bioinformatics tools can help identify these mechanisms and suggest strategies for overcoming them.

Genetic Mutations and Pathway Rewiring:

- Whole-exome sequencing (WES) and RNA-Seq can identify mutations that cause resistance to targeted therapies (e.g., **EGFR T790M mutation** in non-small cell lung cancer), leading to the design of second-line inhibitors that specifically target resistant variants.

MicroRNA and Epigenetics:

- **Epigenetic profiling** (e.g., DNA methylation, histone modifications) can uncover how changes in gene expression regulation contribute to resistance.

Bioinformatics tools that analyze these epigenetic modifications can aid in identifying new therapeutic targets that reverse resistance mechanisms.

Single-Cell Sequencing:

- Single-cell RNA-Seq enables the identification of rare subpopulations of resistant cells within a tumor that might be missed in bulk sequencing, providing insights into **heterogeneity** and the evolution of drug resistance.

Clinical Applications of Bioinformatics in Drug Development

- **Drug Development Pipeline Optimization:** Bioinformatics helps to streamline the drug development pipeline by reducing the need for extensive trial-and-error testing. By providing insights into which compounds are likely to interact with the intended target and how they affect tumor behavior, bioinformatics can reduce both the time and cost of drug discovery.
- **Adverse Event Prediction:** Bioinformatics tools can predict potential side effects and adverse reactions to new drugs by analyzing genomic data related to drug metabolism and toxicity. Databases like **PubChem** and **TOXNET** provide information on chemical toxicity and drug interactions, aiding in the design of safer therapies.

Bioinformatics has fundamentally transformed cancer drug development, enabling more targeted, personalized approaches to therapy. Through the integration of genomic, transcriptomic, proteomic, and clinical data, bioinformatics tools are enhancing target discovery, optimizing clinical trial design, and identifying biomarkers for patient stratification. Furthermore, bioinformatics is crucial in understanding the mechanisms of drug resistance and in designing next-generation therapies. As computational methods continue to evolve, bioinformatics will remain at the forefront of accelerating the discovery of new, more effective cancer treatments.

16 Chapter

BIOINFORMATICS APPLICATIONS AND FUTURE DIRECTIONS

INTRODUCTION

Bioinformatics is a multidisciplinary field that merges biology, computer science, and information technology to analyze and interpret complex biological data. As the volume of biological data—particularly genomic, transcriptomic, and proteomic information—continues to grow exponentially, bioinformatics has become essential in deciphering the underlying molecular mechanisms of health, disease, and evolution. By utilizing algorithms, computational models, and database management systems, bioinformatics provides powerful tools for discovering new drug targets, personalizing medical treatments, and advancing our understanding of biological processes. Looking ahead, the future of bioinformatics holds great promise, with emerging technologies such as artificial intelligence, machine learning, and big data analytics poised to further revolutionize the field, driving innovations in precision medicine, biotechnology, and genomics.

APPLICATIONS OF BIOINFORMATICS IN DRUG DISCOVERY

- Bioinformatics plays a crucial role in modern drug discovery, offering innovative tools and methodologies that enhance the efficiency, precision, and success rates of drug development. The process of drug discovery involves identifying potential drug targets, screening compound libraries, and understanding the molecular interactions between drugs and their targets. Traditionally, these steps were time-consuming and costly, but bioinformatics has revolutionized this process by enabling the analysis of vast amounts of biological and chemical data to uncover promising drug candidates more quickly and at a lower cost.
- One of the key applications of bioinformatics in drug discovery is **target identification and validation**. A deep understanding of the molecular basis of diseases is critical for identifying proteins or genes that play pivotal roles in disease mechanisms. Bioinformatics tools, such as gene expression analysis, protein structure prediction, and pathway analysis, allow researchers to pinpoint potential targets with high precision. For

example, through comparative genomics and transcriptomic data analysis, bioinformatics can identify genes that are overexpressed or mutated in specific diseases, such as cancer or neurodegenerative disorders. Furthermore, bioinformatics can help validate these targets by predicting their druggability based on factors like their protein structure, binding sites, and the availability of suitable ligands.

- Once potential drug targets are identified, bioinformatics continues to play a vital role in **drug screening and design**. Virtual screening techniques, powered by computational methods like molecular docking and molecular dynamics simulations, allow researchers to evaluate large libraries of chemical compounds for their potential to bind to a specific target. These in silico approaches are much faster and more cost-effective than traditional high-throughput screening methods, enabling the identification of lead compounds that can be further optimized. Additionally, bioinformatics-driven **drug design** uses structural information, such as 3D protein structures, to design novel small molecules or biologics that can specifically interact with the target. This process, often referred to as *structure-based drug design* (SBDD), helps optimize the binding affinity, specificity, and stability of drug candidates, accelerating the development of new therapeutics.
- In the later stages of drug development, bioinformatics continues to contribute significantly through pharmacogenomics—the study of how genetic variations influence an individual's response to drugs. By analyzing genomic data from diverse populations, bioinformatics tools can identify genetic markers associated with drug metabolism, efficacy, and toxicity. This enables the development of personalized medicine strategies, where drugs are tailored to individuals based on their genetic profiles, improving therapeutic outcomes while minimizing adverse effects. Furthermore, bioinformatics is also applied in toxicology studies to predict the potential toxicity of drug candidates. Computational models can assess the likelihood of a compound causing harmful side effects, helping to prioritize safer candidates for further development and clinical trials.
- The integration of omics data—such as genomics, proteomics, metabolomics, and transcriptomics—into drug discovery is another area where bioinformatics has had a profound impact. These datasets offer a holistic view of biological systems, allowing researchers to understand complex disease mechanisms, identify novel biomarkers, and predict how drugs will interact with multiple cellular pathways. Bioinformatics tools such as network analysis, machine learning, and data mining help researchers integrate and interpret these large, multidimensional datasets,

providing insights that are crucial for the discovery of new drug targets and optimizing drug efficacy.

- Another exciting development in bioinformatics-driven drug discovery is the use of artificial intelligence (AI) and machine learning (ML) algorithms to predict drug properties, optimize compound design, and even suggest novel drug candidates. Machine learning models can analyze vast amounts of biological, chemical, and clinical data to identify patterns and predict the likelihood of a compound's success in clinical trials. AI-based tools are also being used to design new chemical entities that are more likely to bind to targets, exhibit low toxicity, and have favorable pharmacokinetic profiles. The application of AI in drug discovery has the potential to significantly accelerate the entire drug development pipeline, from early-stage discovery to clinical development.
- Finally, bioinformatics is helping to address the growing problem of drug resistance. In diseases like cancer, HIV, and tuberculosis, patients can develop resistance to existing therapies over time. Bioinformatics tools can help identify mutations in the target proteins that confer resistance, enabling the development of second- or third-line drugs that can circumvent these resistance mechanisms. By analyzing genomic data from resistant strains or tumor samples, bioinformatics allows for a better understanding of the molecular evolution of resistance, leading to more effective therapeutic strategies and combination treatments.
- Bioinformatics is transforming the drug discovery process by providing powerful computational tools to identify new drug targets, design and optimize drug candidates, predict drug responses, and assess safety profiles. The continued integration of bioinformatics with emerging technologies like AI, machine learning, and big data analytics promises to accelerate the development of more effective, personalized, and safer therapeutics. As we move forward, bioinformatics will remain at the forefront of drug discovery, driving innovation and improving the efficiency of bringing life-saving drugs to patients worldwide.

BIOINFORMATICS IN AGRICULTURAL GENOMICS

- Bioinformatics has become an essential tool in the field of agricultural genomics, revolutionizing our ability to enhance crop and livestock production, improve disease resistance, and create more sustainable agricultural systems. With the growing global demand for food, driven by population growth and climate change, the need to optimize agricultural productivity and resilience has never been greater. Bioinformatics provides the computational power and analytical frameworks necessary to decode the genetic makeup of plants and animals, uncover valuable traits, and facilitate the development of more robust and high-yielding varieties. Through the integration of genomics, data science, and advanced computational tools, bioinformatics is accelerating breakthroughs in plant breeding, pest resistance, nutrient optimization, and more.

- One of the primary applications of bioinformatics in agricultural genomics is in genome sequencing and annotation. Sequencing the genomes of crops, livestock, and beneficial microorganisms offers a wealth of information about their genetic composition, enabling researchers to identify genes associated with desirable traits such as drought tolerance, pest resistance, or higher nutritional value. In crops, for example, bioinformatics tools are used to assemble and annotate complex plant genomes, many of which are large, polyploid, and contain repetitive sequences. High-throughput sequencing technologies like next-generation sequencing (NGS) generate massive datasets that require advanced computational methods to assemble and interpret. Bioinformatics platforms facilitate the annotation of these genomes, identifying key genes and regulatory elements, thus providing breeders with the molecular targets they need to enhance crop performance. The sequencing of plant and animal genomes is also instrumental in understanding the genetic basis of diseases, enabling the development of disease-resistant varieties or breeds.
- In the context of plant breeding, bioinformatics has enabled the rise of genomic selection, a cutting-edge technique that accelerates the development of new crop varieties with desirable traits. Genomic selection involves using molecular markers, derived from genomic data, to predict the breeding values of individual plants or animals. This method allows breeders to select the best candidates early in the breeding process, long before they mature or flower. Bioinformatics tools are used to identify and catalog the genetic markers associated with traits like yield, pest resistance, and stress tolerance. By analyzing the entire genome and its relationship with phenotypic data, bioinformatics platforms can predict the potential of different crosses, enabling faster and more efficient breeding programs. This is particularly important in crops that take several seasons to mature, as genomic selection can significantly reduce the time and resources required to bring new, improved varieties to market.
- Marker-assisted selection (MAS) is another key area where bioinformatics has made a significant impact in agriculture. MAS uses genetic markers to select individuals that possess desirable traits without needing to

evaluate the physical expression of these traits in the plant or animal. For example, in rice, maize, and wheat, bioinformatics-driven identification of molecular markers linked to traits such as drought tolerance, high yield, or disease resistance has helped accelerate the breeding of new varieties. Bioinformatics tools analyze data from genome-wide association studies (GWAS) and quantitative trait loci (QTL) mapping to identify genetic variants that correlate with specific traits. These markers are then used in breeding programs to select for these traits more efficiently. This allows for the precise and targeted improvement of crop varieties, reducing the reliance on traditional, time-intensive breeding methods.

- In livestock genomics, bioinformatics is transforming breeding strategies by enabling the identification of genetic markers for traits such as growth rate, disease resistance, milk production, and reproductive efficiency. The genetic diversity within livestock populations is vast, and bioinformatics tools are crucial for understanding this complexity. With whole-genome sequencing and SNP (single nucleotide polymorphism) analysis, researchers can identify beneficial alleles in livestock breeds and use this information to enhance breeding programs. For example, genomic selection in dairy cattle has led to improvements in milk yield, disease resistance, and overall herd health. Similarly, bioinformatics tools are used to study the genomes of poultry, pigs, and other livestock species to improve feed efficiency, disease resistance, and growth rates, all while ensuring ethical and sustainable practices in animal breeding.
- Another vital application of bioinformatics in agricultural genomics is in pest and disease resistance. One of the greatest challenges in agriculture is protecting crops and livestock from a wide array of pests, pathogens, and diseases. Bioinformatics helps by enabling the genetic mapping of resistance traits, which can then be incorporated into breeding programs. For example, bioinformatics tools have been used to identify resistance genes in crops such as rice, wheat, and tomatoes, which are essential for developing varieties that can withstand common diseases like blight, rust, or bacterial wilt. In addition, by sequencing the genomes of pests, such as insects and fungi, bioinformatics can reveal the genetic mechanisms behind pest resistance to pesticides or herbicides. This knowledge helps design new, more effective pest management strategies and improves the sustainability of agricultural practices by reducing dependence on chemical treatments.
- Metagenomics and microbiome analysis represent another important frontier in agricultural genomics where bioinformatics is making strides.

The microbiome of soil, plants, and animals plays a critical role in plant health, nutrient uptake, and disease resistance. By sequencing and analyzing the genomes of microbes, bioinformatics can provide insights into the composition and function of microbiomes in agricultural systems. For example, soil microbes can influence crop yield and soil fertility, while plant-associated microbes can enhance plant growth and resistance to stress. Bioinformatics platforms are used to analyze large-scale metagenomic data, enabling the identification of beneficial microbial species or strains that can be used as biofertilizers or biopesticides. Additionally, microbiome data can be used to improve livestock health, as gut microbiota plays a crucial role in digestion, immunity, and overall animal performance.

- Bioinformatics is also instrumental in nutritional genomics, where the focus is on enhancing the nutritional quality of crops and livestock. With the help of genomic data, bioinformatics tools are used to identify genes involved in the synthesis of essential nutrients such as vitamins, amino acids, and antioxidants in plants and animals. For example, bioinformatics has been applied to improve the iron and zinc content in staple crops like rice and wheat, addressing global micronutrient deficiencies. In livestock, bioinformatics helps optimize the production of high-quality protein, omega-3 fatty acids, and other beneficial nutrients. The knowledge gained through these efforts is crucial for combating malnutrition and promoting global food security.
- Lastly, systems biology and big data analytics are enabling the integration of diverse biological data sources, such as genomics, transcriptomics, metabolomics, and phenomics, to develop more holistic models of agricultural systems. Bioinformatics tools that can handle and analyze large-scale data from various omics platforms are helping researchers understand the complex interactions between genes, environments, and traits. These models are critical for predicting how crops and livestock will respond to changing environmental conditions, such as climate change, and for optimizing agricultural practices based on predictive insights.
- Bioinformatics is fundamentally transforming agricultural genomics by providing the computational power needed to accelerate crop improvement, enhance livestock breeding, combat pests and diseases, and improve the sustainability and nutritional quality of agricultural products. By integrating large datasets, advanced computational tools, and cutting-edge genomic technologies, bioinformatics is paving the way for more efficient, precise, and sustainable agricultural practices. The future of agricultural genomics is closely tied to the continued evolution of bioinformatics, with

the promise of meeting global food security challenges and addressing the environmental and economic concerns associated with traditional farming practices.

ARTIFICIAL INTELLIGENCE AND MACHINE LEARNING IN BIOINFORMATICS

- Artificial Intelligence (AI) and Machine Learning (ML) have rapidly emerged as transformative forces within bioinformatics, offering revolutionary advancements in how biological data is analyzed, interpreted, and applied. With the sheer volume and complexity of biological data generated in fields like genomics, proteomics, transcriptomics, and systems biology, traditional computational approaches often struggle to keep pace. The integration of AI and ML into bioinformatics is not only enhancing our ability to analyze massive datasets but also enabling predictive modeling, personalized medicine, and the identification of novel drug targets, biomarkers, and disease mechanisms. This paradigm shift has opened up new avenues for scientific discovery, helping researchers uncover previously hidden patterns in complex biological systems that were once beyond our reach.
- One of the most significant contributions of AI and ML to bioinformatics is in the realm of genomic data analysis. The explosion of genomic sequencing technologies, such as next-generation sequencing (NGS) and single-cell RNA sequencing, has generated vast amounts of data that require advanced computational methods for interpretation. ML algorithms, particularly deep learning models, are used to analyze large-scale genomic datasets, identifying genetic variants, mutations, and structural variations associated with diseases such as cancer, neurological disorders, and cardiovascular diseases. AI tools can automatically identify regions of the genome that are of biological significance, facilitating faster and more accurate genome-wide association studies (GWAS). For instance, convolutional neural networks (CNNs), a form of deep learning, are employed to detect patterns in genomic sequences, such as promoter regions, enhancers, and other regulatory elements that may contribute to disease. In this way, AI not only speeds up the analysis of complex genomic data but also improves the sensitivity and specificity of genetic variant detection, which is crucial for understanding the molecular basis of diseases and identifying potential therapeutic targets.
- In protein structure prediction, AI and ML are making significant strides, particularly with tools like AlphaFold from DeepMind, which have revolutionized our understanding of protein folding. Protein structure is fundamental to understanding its function and interactions within the cell,

and misfolded proteins are often implicated in diseases like Alzheimer's, Parkinson's, and cystic fibrosis. Predicting the three-dimensional structure of proteins from their amino acid sequences has long been a challenge in structural bioinformatics, but machine learning techniques, especially deep learning, have transformed this task. AlphaFold and similar AI-based models can now predict protein structures with remarkable accuracy, even in cases where experimental data is scarce or unavailable. This advancement opens up new possibilities for drug discovery, as understanding protein structures can help in the design of small molecules or biologics that can specifically target disease-causing proteins.

- In drug discovery and development, AI and ML are also having a profound impact, accelerating the identification of novel drug candidates, optimizing drug design, and predicting drug efficacy and toxicity. Traditional drug discovery methods are time-consuming and expensive, often requiring years of research and billions of dollars in investment. However, AI-driven tools can predict how molecules will interact with biological targets by analyzing large datasets of chemical, biological, and clinical information. Machine learning models can be trained to identify potential drug candidates by analyzing known chemical structures and their effects on biological systems. For example, deep learning models can analyze vast compound libraries and predict which compounds are most likely to bind to a specific target, dramatically reducing the need for high-throughput screening. These predictive models are particularly useful for identifying novel drug candidates for hard-to-target diseases, such as neurodegenerative disorders or rare cancers. Furthermore, ML algorithms can also predict drug toxicity, side effects, and optimal dosing, enabling safer and more effective drug development.
- AI and ML have also made significant contributions to personalized medicine, where treatments are tailored to an individual's genetic profile, lifestyle, and environment. In cancer treatment, for example, AI models can analyze genomic data from patients' tumors to predict the most effective therapy based on the genetic mutations present. By integrating genomic, transcriptomic, and clinical data, machine learning algorithms can generate personalized treatment plans that maximize efficacy and minimize adverse effects. In addition to genetic information, ML models can integrate other omics data (e.g., proteomics, metabolomics) and real-time clinical data to monitor treatment response and predict patient outcomes. This integration of diverse data types is essential for advancing precision medicine, as it allows for more accurate predictions of how patients will

respond to treatments, thereby optimizing therapeutic strategies and improving clinical outcomes.

- The application of AI and ML in biomarker discovery has also led to groundbreaking developments. Identifying biomarkers for disease detection, prognosis, and treatment response is one of the primary challenges in modern medicine. Machine learning algorithms excel in identifying correlations and patterns in complex biological datasets, which is essential for discovering new biomarkers. For example, in cancer research, AI models can analyze genomic, transcriptomic, and imaging data to identify novel biomarkers that predict disease progression or response to treatment. Similarly, AI-driven algorithms can identify microbial biomarkers that serve as indicators of disease in microbiome research. Once discovered, these biomarkers can be used in diagnostic tests, enabling earlier and more accurate detection of diseases, as well as the development of targeted therapies.
- The integration of big data analytics and AI-driven systems biology is another area where bioinformatics is seeing tremendous growth. Biological systems are complex and interconnected, with numerous factors influencing gene expression, protein activity, metabolic pathways, and cellular functions. AI and ML algorithms, such as unsupervised learning and reinforcement learning, can analyze vast networks of biological data to model these systems and predict how they will respond to various stimuli, including drugs, environmental changes, or genetic mutations. This approach is especially powerful in understanding complex diseases like cancer, diabetes, and neurodegenerative diseases, where multiple factors contribute to disease progression. By simulating these systems and predicting how biological networks will behave under different conditions, AI can help researchers design more effective therapeutic strategies and identify novel drug targets.
- Furthermore, AI and ML are playing an important role in epidemiology and public health. Machine learning algorithms can analyze large-scale genomic and clinical data to identify patterns of disease transmission, predict outbreaks, and assess the effectiveness of public health interventions. For example, during the COVID-19 pandemic, AI-driven models were used to predict disease spread, identify potential hotspots, and optimize the allocation of resources. In global health, AI can be leveraged to track genetic variants of pathogens, such as viruses or bacteria, to monitor their evolution and predict future outbreaks. This ability to analyze and interpret large datasets quickly and accurately is essential for responding to emerging infectious diseases and other public health challenges.

- One of the challenges in applying AI and ML to bioinformatics is the need for data quality and standardization. Biological data is often noisy, incomplete, or biased, and the effectiveness of AI models depends on the quality of the data used to train them. Ensuring that datasets are accurate, representative, and properly annotated is crucial for the successful application of AI in bioinformatics. Moreover, the ethical considerations surrounding the use of AI in healthcare, such as data privacy, algorithmic transparency, and fairness, must also be addressed to ensure that these technologies benefit society as a whole.
- AI and ML are reshaping bioinformatics by providing powerful tools to analyze complex biological data, uncover hidden patterns, and predict outcomes. From genome analysis and drug discovery to personalized medicine and systems biology, AI-driven approaches are accelerating scientific discoveries and improving healthcare. As computational power continues to grow and data availability expands, the integration of AI and ML in bioinformatics will only become more pervasive, unlocking new frontiers in biological research and clinical application. The future of bioinformatics lies in the continued evolution of AI technologies, which promise to revolutionize our understanding of biology and transform the way we approach health and disease.

REFERENCES

Alekseyev, M. A. and P. A. Pevzner (2009). "Breakpoint graphs and ancestral genome reconstructions". Genome Research Vol. 19: 943–957.

Altschul S, Madden T, Schaffer A, et al. Gapped BLAST and PSI-BLAST: a new generation of protein database search programs. Nucleic Acids Res. 1997;25:3389–402.

Ane C, Larget B, Baum D, Smith S, Rokas A. Bayesian estimation of concordance among gene trees. Mol Biol Evol. 2007;24:412–26.

Anjar U. Scibuntu: Ubuntu Linux for scientists. Version 0.4-beta. 2006; Available from http://scibuntu.sourceforge.net/index.html.

Bafna, V. and P. A. Pevzner (1996). "Genome Rearrangements and Sorting by Reversals". SIAM Journal on Computing Vol. 25: 272–289.

Bioinformatics and Functional Genomics by Jonathan Pevsner. YEAR Publisher WileyBlackwell

Bortolussi N, Durand E, Blum M, Francois O. apTreeshape: statistical analysis of phylogenetic tree shape. Bioinformatics. 2006;22:363–4.

Butler MA, King AA. Phylogenetic comparative analysis: A modeling approach for adaptive evolution. Am Nat. 2004;164:683–95.

Butler, J., I MacCallum, M. Kleber, I. A. Shlyakhter, M. Belmonte, E. S. Lander, C. Nusbaum, and D. B. Jaffe (2008). "ALLPATHS: de novo assembly of whole-genome shotgun microreads." Genome Research Vol. 18: 810–820.

Chapman B, Chang J. Biopython: Python tools for computational biology. ACM SIGBIO Newsletter. 2000;20:15–9.

Chenna R, Sugawara H, Koike T, et al. Multiple sequence alignment with the Clustal series of programs. Nucleic Acids Res. 2003;31:3497–500.

Chessel D, Dufour AB, Thioulouse J. The ADE4 package I: One-table methods. R News. 2004;4:5–10.

Cohen, D. S. and M. Blum (1995). "On the Problem of Sorting Burnt Pancakes". Discrete Applied Mathematics Vol. 61: 105–120.

Conti, E., N. P. Franks, and P. Brick (1996). "Crystal structure of firefly luciferase throws light on a superfamily of adenylate-forming enzymes". Structure Vol. 4: 287–298.

Conti, E., T. Stachelhaus, M. A. Marahiel, and P. Brick (1997). "Structural basis for the activation of phenylalanine in the non-ribosomal biosynthesis of gramicidin S". The EMBO Journal Vol. 16: 4174–4183.

Cristianini, N. and M. W. Hahn (2006). Introduction to Computational Genomics: A Case Studies Approach. Cambridge University Press.

de Bruijn, N. (1946). "A Combinatorial Problem". In: Proceedings of the Section of Sciences, Koninklijke Akademie van Wetenschappen te Amsterdam. Vol. 49: 758–764.

Doolittle, R. F., M. W. Hunkapiller, L. E. Hood, S. G. Devare, K. C. Robbins, S. A. Aaronson, and H. N. Antoniades (1983). "Simian sarcoma virus onc gene, v-sis, is derived from the gene (or genes) encoding a platelet-derived growth factor". Science Vol. 221: 275–277.

Drmanac, R., I. Labat, I. Brukner, and R. Crkvenjakov (1989). "Sequencing of megabase plus DNA by hybridization: Theory of the method". Genomics Vol. 4: 114–128.

Drummond AJ, Rambaut A. BEAST: Bayesian evolutionary analysis by sampling trees. BMC Evol Biol. 2007;7:214–22.

Eddy SR. Profile hidden Markov models. Bioinformatics. 1998;14:755–63.

Edgar RC. MUSCLE: Multiple sequence alignment with high accuracy and high throughput. Nucleic Acids Res. 2004;32:1792–7.

Euler, L. (1758). "Solutio Problematis ad Geometriam Situs Pertinentis". Novi Commentarii Academiae Scientarium Imperialis Petropolitanque, 9–28.

Evolution by Nicholas H Barton, Derek E.G. Briggs, Jonathan A Eisen, David B. Goldstein, Nipam H Patel.2007-2010. Publisher CSHL press

Felsenstein J. PHYLIP (phylogeny inference package). Version 3.68. 2009; Available from http://evolution.genetics.washington.edu/phylip.html.

Galtier N, Gouy M, Gautier C. SEAVIEW and PHYLO_WIN: two graphic tools for sequence alignment and molecular phylogeny. Bioinformatics. 1996;12:543–8.

Gao, F. and C.-T. Zhang (2008). "Ori-Finder: A web-based system for finding oriCs in unannotated bacterial genomes". BMC Bioinformatics Vol. 9: 79.

Gardner, M. (1974). "Mathematical Games". Scientific American Vol. 230: 120–125.

Gates, W. H. and C. H. Papadimitriou (1979). "Bounds for sorting by prefix reversal". Discrete Mathematics Vol. 27: 47–57.

Geman, S. and D. Geman (1984). "Stochastic Relaxation, Gibbs Distributions, and the Bayesian Restoration of Images". IEEE Transactions on Pattern Analysis and Machine Intelligence Vol. PAMI-6: 721–741.

Gilbert D. Readseq. Version 2. 2001; Available from http://iubio.bio.indiana. edu/soft/molbio/readseq/java/.

Good, I. J. (1946). "Normal Recurring Decimals". Journal of the London Mathematical Society Vol. 21: 167–169.

Grasso C, Lee C. Combining partial order alignment and progressive multiple sequence alignment increases alignment speed and scalability to very large alignment problems. Bioinformatics. 2004;20:1546–56.

Grigoriev, A. (1998). "Analyzing genomes with cumulative skew diagrams". Nucleic Acids Research Vol. 26: 2286–2290.

Grigoriev, A. (2011). "How do replication and transcription change genomes?" In: Bioinformatics for Biologists. Ed. by P. A. Pevzner and R. Shamir. Cambridge University Press, 111–125.

Guibas, L. and A. Odlyzko (1981). "String overlaps, pattern matching, and nontransitive games". Journal of Combinatorial Theory, Series A Vol. 30: 183 –208.

Guindon S, Gascuel O. A simple, fast, and accurate algorithm to estimate large phylogenies by maximum likelihood. Syst Biol. 2003;52:696–704.

Hannenhalli, S. and P. A. Pevzner (1999). "Transforming Cabbage into Turnip: Polynomial Algorithm for Sorting Signed Permutations by Reversals". Journal of the ACM Vol. 46: 1–27.

Harmer, S. L., J. B. Hogenesch, M. Straume, H. S. Chang, B. Han, T. Zhu, X. Wang, J. A. Kreps, and S. A. Kay (2000). "Orchestrated transcription of key pathways in Arabidopsis by the circadian clock". Science Vol. 290: 2110–2113.

Harmon LJ, Weir JT, Brock CD, Glor RE, Challenger W. GEIGER: investigating evolutionary radiations. Bioinformatics. 2008;24:129–31.

Hertz, G. Z. and G. D. Stormo (1999). "Identifying DNA and protein patterns with statistically significant alignments of multiple sequences." Bioinformatics Vol. 15: 563–577.

Hillis DM. The tree of life and the grand synthesis of biology. In: Cracraft J, Donoghue MJ, eds. Assembling the tree of life. New York: Oxford University Press; 2004:545–7.

Idury, R. M. and M. S. Waterman (1995). "A New Algorithm for DNA Sequence Assembly." Journal of Computational Biology Vol. 2: 291–306.

Ivics, Z., P. Hackett, R. Plasterk, and Z. Izsvák (1997). "Molecular reconstruction of Sleeping Beauty, a Tc1-like transposon from fish, and its transposition in human cells". Cell Vol. 91: 501–510.

Katoh K, Misawa K, Kuma K, Miyata T. MAFFT: a novel method for rapid multiple sequence alignment based on fast Fourier transform. Nucleic Acids Res. 2002;30:3059–66.

Konopka, R. J. and S. Benzer (1971). "Clock mutants of Drosophila melanogaster". Proceedings of the National Academy of Sciences of the United States of America Vol. 68: 2112–2116.

Lawrence, C. E., S. F. Altschul, M. S. Boguski, J. S. Liu, A. F. Neuwald, and J. C. Wootton (1993). "Detecting subtle sequence signals: a Gibbs sampling strategy for multiple alignment". Science Vol. 262: 208–214.

Levenshtein, V. I. (1966). "Binary codes capable of correcting deletions, insertions, and reversals". Soviet Physics Doklady Vol. 10: 707–710.

Liachko, I., R. A. Youngblood, U. Keich, and M. J. Dunham (2013). "High-resolution mapping, characterization, and optimization of autonomously replicating sequences in yeast". Genome Research Vol. 23: 698–704.

Liu L, Pearl DK. Species trees from gene trees: Reconstructing Bayesian posterior distributions of a species phylogeny using estimated gene tree distributions. Syst Biol. 2007;56:504–14.

Lobry, J. R. (1996). "Asymmetric substitution patterns in the two DNA strands of bacteria". Molecular Biology and Evolution Vol. 13: 660–665.

Lundgren, M., A. Andersson, L. Chen, P. Nilsson, and R. Bernander (2004). "Three replication origins in Sulfolobus species: Synchronous initiation of chromosome replication and asynchronous termination". Proceedings of the National Academy of Sciences of the United States of America Vol. 101: 7046–7051.

Lysov, Y., V. Florent'ev, A. Khorlin, K. Khrapko, V. Shik, and A. Mirzabekov (1988). "DNA sequencing by hybridization with oligonucleotides". Doklady Academy Nauk USSR Vol. 303: 1508–1511.

Ma, J., A. Ratan, B. J. Raney, B. B. Suh, W. Miller, and D. Haussler (2008). "The infinite sites model of genome evolution". Proceedings of the National Academy of Sciences of the United States of America Vol. 105: 14254–14261.

Maddison W, Maddison D. Mesquite: a modular system for evolutionary analysis. Version 2.6. 2009; Available from http://mesquiteproject.org.

Mahajan BK 2002 (Methods in Biostatistics) (6th edition)

Maxam, A. M. and W. Gilbert (1977). "A new method for sequencing DNA." Proceedings of the National Academy of Sciences of the United States of America Vol. 74: 560–564.

Medvedev, P., S. K. Pham, M. Chaisson, G. Tesler, and P. A. Pevzner (2011). "Paired de Bruijn Graphs: A Novel Approach for Incorporating Mate Pair Information into Genome Assemblers." Journal of Computational Biology Vol. 18: 1625-1634.

Morgenstern B. DIALIGN 2: Improvement of the segment-to-segment approach to multiple sequence alignment. Bioinformatics. 1999;15:211–8.

Nadeau, J. H. and B. A. Taylor (1984). "Lengths of chromosomal segments conserved since divergence of man and mouse". Proceedings of the National Academy of Sciences of the United States of America Vol. 81: 814–818.

NERC Environmental Bioinformatics Centre. Bio-Linux. Version 5.0. 2009; Available from http://nebc.nox.ac.uk/tools/bio-linux.

Ng, J., N. Bandeira, W.-T. Liu, M. Ghassemian, T. L. Simmons, W. H. Gerwick, R. Linington, P. Dorrestein, and P. A. Pevzner (2009). "Dereplication and de novo sequencing of nonribosomal peptides". Nature Methods Vol. 6: 596–599.

Notredame C, Higgins DG, Heringa J. T-coffee: A novel method for fastand accurate multiple sequence alignment. J Mol Biol. 2000;302:205–17.

Ohno, S. (1973). "Ancient linkage groups and frozen accidents". Nature Vol. 244: 259- 262.

Paradis E, Claude J, Strimmer K. APE: Analyses of Phylogenetics and Evolution in R language. Bioinformatics. 2004;20:289–90.

Park, H. D., K. M. Guinn, M. I. Harrell, R. Liao, M. I. Voskuil, M. Tompa, G. K. Schoolnik, and D. R. Sherman (2003). "Rv3133c/dosR is a transcription factor that mediates the hypoxic response of Mycobacterium tuberculosis". Molecular Microbiology Vol. 48: 833–843.

Pevzner, P. A. (1989). "1-Tuple DNA sequencing: computer analysis". Journal of Biomolecular Structure and Dynamics Vol. 7: 63–73.

Pevzner, P. A., H Tang, and M. S. Waterman (2001). "An Eulerian path approach to DNA fragment assembly". Proceedings of the National Academy of Sciences of the United States of America Vol. 98: 9748–53.

Pevzner, P. and G. Tesler (2003a). "Genome rearrangements in mammalian evolution: lessons from human and mouse genomes". Genome Research Vol. 13: 37–45.

Pevzner, P. and G. Tesler (2003b). "Human and mouse genomic sequences reveal extensive breakpoint reuse in mammalian evolution". Proceedings of the National Academy of Sciences of the United States of America Vol. 100: 7672–7677.

Posada D. jModelTest: Phylogenetic model averaging. Mol Biol Evol. 2008;25:1253–6.

Rambaut A, Grass NC. Seq-Gen: an application for the Monte Carlo simulation of DNA sequence evolution along phylogenetic trees. Bioinformatics. 1997;13:235–8.

Ronquist F, Huelsenbeck JP. MrBayes 3: Bayesian phylogenetic inference under mixed models. Bioinformatics. 2003;19:1572–4.

Rosenblatt, J. and P. Seymour (1982). "The Structure of Homometric Sets". SIAM Journal on Algebraic Discrete Methods Vol. 3: 343–350.

Sanderson MJ. Paloverde: an OpenGL 3D phylogeny browser. Bioinformatics. 2006;22:1004–6.

Sanderson MJ. r8s: inferring absolute rates of molecular evolution and divergence times in the absence of a molecular clock. Bioinformatics. 2003 2003;19:301–2.

Sanger, F, S Nicklen, and A. Coulson (1977). "DNA sequencing with chain-terminating inhibitors". Proceedings of The National Academy of Sciences of The United States Of America Vol. 74: 5463–5467.

Sedgewick, R. and P. Flajolet (2013). An Introduction to the Analysis of Algorithms. Addison Wesley. 352.

Sernova, N. V. and M. S. Gelfand (2008). "Identification of replication origins in prokaryotic genomes". Briefings in Bioinformatics Vol. 9: 376–391.

Smith SA, Dunn CW. Phyutility: a phyloinformatics tool for trees, alignments and molecular data. Bioinformatics. 2008;24:715–6.

Smith, T. F. and M. S. Waterman (1981). "Identification of common molecular subsequences." Journal of Molecular Biology Vol. 147: 195–197.

Solov'ev, A. (1966). "A combinatorial identity and its application to the problem about the first occurence of a rare event". Theory of Probability and its Applications Vol. 11: 276–282.

Southern, E. (1988). "Analysing Polynucleotide Sequences". Patent (United Kingdom).

Stachelhaus, T., H. D. Mootz, and M. A. Marahiel (1999). "The specificity-conferring code of adenylation domains in nonribosomal peptide synthetases." Chemistry & Biology Vol. 6: 493–505.

Stajich J, Block D, Boulez K, et al. The Bioperl toolkit: Perl modules for the life sciences. Genome Res. 2002;12:1611–8.

Stamatakis A. RAxML-VI-HPC: Maximum likelihood-based phylogenetic analyses with thousands of taxa and mixed models. Bioinformatics. 2006;22:2688.

Sturtevant, A. H. (1921). "A Case of Rearrangement of Genes in Drosophila". Proceedings of the National Academy of Sciences of the United States of America Vol. 7: 235–237.

Sturtevant, A. H. and T. Dobzhansky (1936). "Inversions in the Third Chromosome of Wild Races of Drosophila Pseudoobscura, and Their Use in the Study of the History of the Species". Proceedings of the National Academy of Sciences of the United States of America Vol. 22: 448–450.

Tang, Y. Q., J. Yuan, G. Osapay, K. Osapay, D. Tran, C. J. Miller, A. J. Ouellette, and M. E. Selsted (1999). "A cyclic antimicrobial peptide produced in primate leukocytes by the ligation of two truncated alpha-defensins". Science Vol. 286: 498–502.

Venkataraman, N., A. L. Cole, P. Ruchala, A. J. Waring, R. I. Lehrer, O. Stuchlik, J. Pohl, and A. M. Cole (2009). "Reawakening retrocyclins: ancestral human defensins active against HIV-1". PLoS Biology Vol. 7: e95.

Wang, X., C. Lesterlin, R. Reyes-Lamothe, G. Ball, and D. J. Sherratt (2011). "Replication and segregation of an Escherichia coli chromosome with two replication origins". Proceedings of the National Academy of Sciences Vol. 108: E243–E250.

Webb C, Ackerly D, Kembel S. Phylocom: software for the analysis of phylogenetic community structure and trait evolution. Bioinformatics. 2008;24:2098–100.

Weir J, Schluter D. PhySim: phylogenetic tree simulation package. Version 1.0. 2007; Available from http://cran.r-project.org/web/packages/PhySim/index.html.

Xia, X. (2012). "DNA replication and strand asymmetry in prokaryotic and mitochondrial genomes". Current Genomics Vol. 13: 16–27.

Yancopoulos, S., O. Attie, and R. Friedberg (2005). "Efficient sorting of genomic permutations by translocation, inversion and block interchange". Bioinformatics Vol. 21: 3340–3346.

Zaman SM, HK Rahim and M Howlader 1982. (Simple Lessons from Biometry), BRRI Research methodology methods and techniques, C. R, Kothari, New Age International Limited Publisher,New Delhi.

Zerbino, D. R. and E. Birney (2008). "Velvet: Algorithms for de novo short read assembly using de Bruijn graphs". Genome Research Vol. 18: 821–829.

Zhao, H. and G. Bourque (2009). "Recovering genome rearrangements in the mammalian phylogeny". Genome Research Vol. 19: 934–942.

INDEX

F

G

H

L

M

N

O

P

www.ingramcontent.com/pod-product-compliance
Lightning Source LLC
LaVergne TN
LVHW062117200926
848484LV00007B/1695